T0207062

The world's water resources are coming under increasing stress, a stress that will become critical globally sometime during the next century. This is due to the rapidly rising population demanding more and more water and an increasing level of affluence. The book discusses the background to this issue and the measures to be taken over the next 20–30 years to overcome some of the difficulties that can be foreseen, and the means of avoiding others, such as the hazard of floods. It looks at the water resource and its assessment and management in an integrated fashion. It deals with the requirements of agriculture and of rural and urban societies and to a lesser extent with those of industry and power, against the background of the needs of the natural environment. It presents a number of ways and means of improving the management of national and international affairs involving fresh water. It highlights the importance of fresh water as a major issue for the environment and for development.

GLOBAL WATER RESOURCE ISSUES

GLOBAL WATER RESOURCE ISSUES

GORDON J. YOUNG
Wilfrid Laurier University, Ontario

JAMES C.I. DOOGE
University College Dublin

JOHN C. RODDA
World Meteorological Organization, Geneva

PUBLISHED BY THE PRESS SYNDICATE OF THE UNIVERSITY OF CAMBRIDGE
The Pitt Building, Trumpington Street, Cambridge, United Kingdom

CAMBRIDGE UNIVERSITY PRESS
The Edinburgh Building, Cambridge CB2 2RU, UK
40 West 20th Street, New York NY 10011–4211, USA
477 Williamstown Road, Port Melbourne, VIC 3207, Australia
Ruiz de Alarcón 13, 28014 Madrid, Spain
Dock House, The Waterfront, Cape Town 8001, South Africa

http://www.cambridge.org

First published 1994
First paperback edition 2003

A catalogue record for this book is available from the British Library

Library of Congress cataloguing in publication data

Young, G. J. (Gordon James)
Global water resource issues / by Gordon J. Young, James C. I. Dooge, John C. Rodda.
p. cm.
Includes bibliographical references and index.
ISBN 0 521 46153 7 hardback
1. Water-supply. 2. Water use. 3. Water quality.
I. Dooge, James. II. Rodda, J.C. III. Title.
TD345.Y68 1994
628.1–dc20 93-38146 CIP

ISBN 0 521 46153 7 hardback
ISBN 0 521 46712 8 paperback

Contents

Foreword

Water, and particularly fresh water, is a matter that must concern every man, woman and child on this planet. Those who today have a wholesome and reliable supply must be anxious to safeguard it against the uncertainties of the changing world. Those with inadequate water aspire to their own supply, as one essential step towards the betterment of their standard of living. Every one of us must also be concerned about the water needed for producing the food we eat, the energy we consume and all the other goods and services that are dependent on this vital liquid.

Water is also an essential ingredient of the environment. Through the hydrological cycle that powers the movement of materials and energy about the earth, water is essential to the existence of virtually every living organism. Since the start of history, human activities have been altering the hydrological cycle and reshaping the face of the earth. Now these modifications are becoming more massive. To meet the water demands of the future, they will distort the natural environment even further. Water, then, encapsulates the quintessential dilemma: development **or** environment. Or can it be development **and** environment?

Global Water Resource Issues sets out the global perspective on water as viewed through the eyes of those who prepared for the United Nations Conference on Environment and Development. I commend it to everyone concerned with this precious commodity – which means all of us.

G.O.P Obasi
Secretary-General
World Meteorological Organization

Preface

The world of the late twentieth century is a place subject to many dynamic forces. Change is evidenced everywhere and in many different ways. Global population has expanded remarkably, as a result, in large measure, of better control of disease and the resulting increase in average lifespan. Population growth has, however, been uneven, being overwhelming in many developing countries and being particularly concentrated in large cities which are growing much faster than their rural surroundings due to migration. Resource depletion is in most places of very great concern. Soil, the basis for much of our food production is being washed away into the ocean at unprecedented rates. Forest resources are being depleted. Mining of fossil fuels for energy production, and of minerals for industrial and agricultural activity, proceeds apace, at rates which cannot be maintained forever. Pollution of the natural environment from a variety of sources is, in many parts of the world, out of control and is undermining the quality of life and often threatening life itself.

As the population expands and as living standards rise, albeit unequally in different parts of the world, so the demands for increased exploitation of natural resources rise. These increasing demands, concentrated particularly in less developed countries, and inextricably intertwined with the desire for industrial and agricultural development, come at a time of raised consciousness of the fundamental importance of conservation of the natural environment.

This modern form of environmental consciousness has evolved primarily in the more-developed countries, in different ways within the activist and decision-making communities. It has evolved in countries which, on the whole, are able to afford the costs of preserving, protecting and restoring the natural environment. Less-developed countries tend to be much more concerned with the immediate desires for economic development and improved standards of living than with the need to be environmentally conscious.

Thus a global debate is under way, matching the thrust for protection of the environment, primarily the view from the developed world, with the desire for economic growth, the perspective typical of many less developed countries. Harmonizing the very different perspectives is one of the greatest challenges for the world community at the moment.

Unfortunately, this debate is taking place against an extremely complex geo-political background. There are many elements to this background. There are, perhaps, three which are of crucial importance. First, although there have been no truly global wars since World War II, there have been literally dozens of local or regional conflicts, ranging from short-lived strife as in the case of the Gulf War to the far longer-term warfare raging in the Horn of Africa. Results of such warfare are incompatible with environmental protection. Economic development for people in such areas is a frustrating dream. Second, there are changes in the world geo-political order of enormous consequence. The recent collapse of the former Soviet Union has led to fundamental realignments in the politics and economics of eastern Europe and to a complete change in the world military balance and may lead to changes in the direction of flows of aid from the richest countries to central and eastern Europe rather than to the least-developed countries. Third, an element which remains ever present, the unwillingness of national politicians to jeopardize their own power and authority by putting global concerns, often with concomitant commitment of their nation's resources, before the short-term interests of their own countries.

Over the past two decades there have been three events which, above all others, have focused and guided world attention in the environmental arena. The first was the United Nations Conference on the Human Environment in 1972 held in Stockholm, Sweden. From this conference there emerged the United Nations Environment Programme and the blossoming of global awareness of the importance of environmental concerns. The second was the publication in 1987 of *Our Common Future*, the report of the World Commission on Environment and Development. The thrust of this report was to show how the environment should not be treated in isolation from economic concerns. It stressed the necessity for the holistic approach as political, economic and environmental concerns are so interdependent. The third event, arising directly from the others, was the United Nations Conference on Environment and Development (UNCED) in June 1992 held in Rio de Janeiro, Brazil. The true value of this extremely ambitious conference can only be properly judged with a longer historical perspective. It is safe to claim, however, that it has already accomplished much by focusing the attention of the world on some of the more pressing issues of our time.

While all environmental concerns are of importance, there is a growing realization that water is the most important common element in environment and development issues. In a very real sense, water is life. Life on Earth started in water and without water life as we know it could not continue. Water transports nutrients and chemicals within the biosphere allowing life in all its forms to continue. Plants and animals alike have a fundamental dependence on water. Water also moulds the surface of the Earth, eroding materials from some localities, transporting it and depositing it in others. Water is the key to the global energy budget, helping redistribute heat from one part of the earth to another. It is the main greenhouse gas with a role that is relatively unknown in controlling the energy balance.

Water is also a key component of socio-economic systems. It is the basis for agriculture, essential for many industries and for much energy production. Its importance for human health and welfare is critical and thus the supply of water for drinking and sanitation is of major concern everywhere. The overabundance of water, leading to catastrophic floods, can be just as damaging as water scarcity leading to drought and famine.

Just as environmental issues must be viewed in a holistic manner, so water issues have to be tackled in an integrated fashion and the linkages with other environmental issues set out. One example is the management of water and land, which must be integrated. Water management, however, necessitates the involvement of parties at different levels in the political hierarchy. Many issues can and should be tackled at a very local level, within communities. Other problems of apportionment and use must be tackled on the provincial or national level, while others, particularly in the case of trans-boundary water bodies, must be the subject of international negotiation.

It was against this background of the underlying importance of fresh water to both the natural environment and to human development that the International Conference on Water and the Environment (ICWE) was convened in Dublin, Ireland, January 1992. ICWE was the most comprehensive water conference since the United Nations Water Conference in Mar del Plata, Argentina in 1977. It attracted more than five hundred participants from 114 countries, 28 United Nations agencies and bodies and 58 non-governmental and inter-governmental organizations. The conference considered a very wide range of water management issues, applicable globally, and commended its findings and recommendations to UNCED. ICWE was a meeting of experts rather than government delegates, convened as the official lead in freshwater issues into the political process of UNCED.

ICWE had several principal objectives:

(a) To assess the current status of the world's freshwater resources in relation to present and future water demands and to identify priority issues for the 1990s;
(b) To develop coordinated inter-sectoral approaches towards managing these resources by strengthening the linkages between the various water programmes;
(c) To formulate environmentally-sustainable strategies and action programmes for the 1990s and beyond to be presented to UNCED;
(d) To bring the above issues, strategies and action programmes to the attention of govern-

ments as a basis for national action programmes and to increase awareness of the environmental consequences and development opportunities in improving the management of water resources.

The structure of debate at ICWE was, in large measure, dictated by the series of Preparatory Committee meetings called in the two years prior to UNCED. In order to be of maximum relevance and usefulness to the formulation of the UNCED recommendations, specifically those for the effective conduct of global environmental and developmental policies, the deliberations at ICWE were structured according to the same breakdown of the issues used in UNCED 'Agenda 21'.

The output from ICWE is contained in a five-page *Dublin Statement* and a forty-page *Report of the Conference*. These are very succinct documents, of necessity brief and to the point. It was remarkable that they could have been put together during the five days of the Conference. Clearly much thought had to be put into organizing the Conference, including the preparation of background material for discussion, presented in accordance with the pre-planned conference format. This background documentation contains a great deal of very valuable information which is worth synthesizing in some detail.

While the aim of ICWE was to cover all major water management issues and to consider their application to a wide variety of circumstances, it is recognized that a few important issues were not treated as fully as they might have been. Water for energy production was mentioned only briefly and the special problems associated with large dams were only lightly treated. The problems of floods and droughts, while alluded to, probably deserve more attention than they were given in Dublin. Trans-boundary water problems, in many places the reason for conflict and highly politically contentious, were purposely not dealt with in depth at ICWE as it was essentially a conference of experts in water technology and management rather than of international affairs of a diplomatic nature. Thus, although not completely comprehensive in coverage of major issues, it is felt that the ICWE deliberations covered a sufficiently broad spectrum of concerns and retained the necessary balance to give an extremely useful overview of current world water issues.

Action on the recommendations from ICWE must be undertaken primarily at the country level and below. With this in mind, countries were invited to prepare and submit reports reflecting their particular concerns. The reports were to be structured under the same headings as indicated above. Some 58 such reports were prepared in time for ICWE and were made available during the conference. It is beyond the scope of this book to synthesize the diverse and very valuable information contained in these reports. That may be the subject of another exercise.

The organization of a conference of the magnitude of UNCED is fraught with difficulties. The same difficulties applied, to a lesser extent, to ICWE. The prime difficulty lay in the fact that there are so many players involved, each having a particular perspective on the issues and a particular agenda to achieve. There are enormous differences between countries regarding freshwater resources. Socio-economic and political differences abound. International agencies each have their own *raison d'être* and their own special interests to promote. Non-governmental organizations are often very vocal in pursuing their own goals. The recommendations for action coming out of ICWE and UNCED must be viewed against this background.

Lastly, it must be recognized that while global conferences deal with issues from the global perspectives, much of the action recommended, especially in the freshwater arena, must be enacted at the local level. The political agenda for action must be agreed to at international and national levels but the action itself must be taken locally.

The main reasons for preparing this book are, then:

(a) To set the ICWE within the spectrum of conferences and meetings leading up to UNCED.

(b) To guide the reader through the mass of documentation prepared for ICWE and to reference important additional relevant documentation.

(c) To explain the principal recommendations for action proposed within the ICWE *Report of the Conference* by expanding on the brief background material presented within that report.

The structure of the book is based on the topics chosen for working group discussions at ICWE which in turn reflect the structure of the Agenda 21 UNCED document. Most of the chapters have an internal structure that reflects the discussion within the working groups and contain brief statements on targets and approximate costs as requested by the UNCED PrepComm (United Nations Conference on Environment and Development, Preparatory Committee). In each case, topics dealt with in the ICWE Report and in Agenda 21 have been summarized in tabular form and the relevant paragraphs cited for ease of reference.

THINK GLOBALLY, ACT LOCALLY.

Acknowledgements

A large number of individuals, institutions, agencies and governments contributed to the International Conference on Water and the Environment (ICWE) and thus to the publication of this volume. Many are listed in Annex 3, while others, those who provided funds, are listed in Annex 4. However, without the support of the executive heads of the agencies most involved, there could have been no Conference and no book. The same must be said for the delegates to the UNCED PrepComms and to the UNCED itself and to their governments, but particularly for the Government of Ireland. Lacking the assistance and involvement of the Department of the Environment in Dublin and particularly of its Secretary, Mr Brendan O'Donoghue, the Conference would have foundered. Equally crucial was the backing of Prof. G.O.P. Obasi (Secretary General of the World Meteorological Organization (WMO)), as was the wholehearted support given by members of that organization, especially Dr D. Axford (Deputy Secretary General), Mr J. Murithi (Director of Administration) and Mr A.W. Kabakibo (Director of Languages, Publications and Conferences Department) together with the members of their departments who provided the numerous services important to the Conference. Also very critical was the help of those members of the Hydrology and Water Resources Department who were not as much involved in ICWE as those named in Annex 5 as members of the Steering Committee.

Much of the material for *Global Water Resource Issues* has been drawn from the documents for the Conference and particularly from the background papers which were produced during the latter part of 1991 by the members of the Intersecretariat Group for Water Resources (ISGWR) listed within Annex 5.

The authors of *Global Water Resource Issues* are very grateful to all these individuals and to the bodies and institutions that have aided and assisted the preparation of this book in their different ways. Their support is gratefully acknowledged.

A water hole in West Africa. This illustration embodies many of the fundamental themes of this book: water as essential for life and development; water as a scarce and often polluted resource in many regions; water supply threatened by changes in climate; the need for improved methods of supply and management; and the role of women in drinking water supply. Credit: WMO.

1
Overview

1.1 Introduction

The International Conference on Water and the Environment, held in Dublin, Ireland, from 26–31 January, 1992, was the most all-embracing conference dealing with global water management issues since the United Nations Water Conference held in Mar del Plata, Argentina in 1977 (United Nations, 1977). Its purpose was to examine current priority issues in the freshwater field and, drawing on expert opinion from a very wide array of nations, institutions and organizations, to recommend the action necessary to alleviate problems. These recommendations were initially taken to the fourth Preparatory Committee for the United Nations Conference on Environment and Development held at the United Nations in New York in March 1992. From there they went to the Earth Summit itself in Rio de Janeiro, Brazil in June 1992 where many of the recommendations were incorporated into the Agenda 21 document, a blueprint for action into the twenty-first century.

The purpose of this book is to enlarge upon and explain the two main documents produced at Dublin, the *Dublin Statement* and the *Report of the Conference* (International Conference on Water and the Environment, 1992). The *Dublin Statement* is reproduced in its entirety in Annex 1, and much of the substance of the *Report of the Conference* is found embedded in the texts of individual chapters of this book.

A theme fundamental to current thinking is that environmental issues in general, and water issues in particular, should not be considered in isolation. To be properly appreciated they must be viewed in as broad a context as possible. This means that water must be examined not only in its role as essential to all life systems on Planet Earth, but also in its role as indispensable to most human activity and thus to socio-economic development. A holistic approach to management of resources and to development issues is now regarded as essential.

Clearly, individual human beings and agglomerations of people living in villages, towns, regions and nation states find themselves in very different circumstances. The circumstances in Bangladesh, where the populace is often threatened by flood and the press of a very high population density, contrast with the almost permanent drought situation of most of sub-Saharan Africa, with its extensive areas of sparse population. Both these situations contrast with the much higher living standards and the far greater security found in the countries of the developed world.

Global realities are constantly and often dramatically changing (see UN–DIESA, World Economic Survey, 1991(a)). World population continues to grow at an alarming rate and at quite different rates in different parts of the globe. The balance of political, military and economic power is in a continuous state of flux. There have been spectacular changes in the recent past in the former Soviet Union and the countries of eastern Europe, leading to a very sudden alteration in geo-political realities. The slower changes in economic power are of fundamental importance too. The rise and relative decline of the USA during the twentieth century has been remarkable. Similarly, the post World War II emergence of Japan and Germany as major economic powers has greatly affected international economic activity. The more recent industrial growth of countries in south east Asia changes the scene once again.

Another fundamental reality is the wide disparity in the consumption patterns between developed and developing countries. Developed countries may be characterized by excessive over-consumption. With few exceptions, people in the highly industrialized economies exhibit profligate and wasteful consumption habits which, in the long run, cannot be sustained. These habits contrast remarkably with the much lower per-capita consumption in the developing world.

There is a need also for a certain historical perspective in order to understand the process leading to ICWE (the International Conference on Water and the Environment) and UNCED (the United Nations Conference on Environment and Development). In the last few decades there has been a quite dramatic and significant change in the philosophy of management. These recent developments in management policy should be examined in order to give the perspective necessary for better understanding.

Thus, this chapter attempts to set the scene by introducing a number of important themes as well as describing the process culminating in ICWE and UNCED.

1.2 The global scene – the overall setting

The purpose of this section is to set water issues in a very wide global context. Water is intimately linked with all aspects of the natural environment and with most human activities. It is therefore relevant to mention some of the more important settings.

1.2.1 The physical setting

Water plays a central role in natural processes at and near to the surface of the Earth, in the atmosphere immediately above the surface and in the soils and rocks immediately below the surface. The quantities of water and the length of residence time of water at any place at or near to the surface is the result of the interaction of climate and the characteristics of the surface.

The climate zones, shown in Figure 1, set the broad pattern for water availability in space and time. Of crucial importance are the total quantities of water available and their variability through time. In any one place or region, variability and reliability are very important. In the mid latitudes there are broad regions where there is relatively little variability from one season to another and where the annual precipitation amounts exhibit relatively small differences from year to year. This contrasts with many arid areas, mostly in the sub-tropics, where there are great differences in precipitation from season to season and where, too, there are often cycles of drought and relative abundance of precipitation over long time periods.

It is now well established that there is a strong likelihood of an overall global warming which is expected to result in significant changes in water resource availability, although detailed regional predictions are at present vague and extremely difficult to determine. It should be noted, however, that while slow, long-term change will almost certainly be of importance, the effects of such change are, for the most part, relatively insignificant compared with the importance of the variability (and hence the unreliability) of existing water supplies. As far as the water manager is concerned, it is the unreliability of the resource which makes his or her job so difficult.

The outlines of the broad climate zones are modified regionally and locally by the general distribution of seas and land masses relative to prevailing wind directions and by the topography of those land masses. Regional and local climates can be greatly modified by the configuration of coastlines relative to rain-bearing winds. Mountain masses are of great regional importance. Their height, lateral extent and orientation greatly influence receipt of precipitation both within their confines and in adjacent regions. Mountains invariably receive far more precipitation than their surrounding areas and experience reduced evaporation; consequently they are important for their water resources. Indeed, mountain regions provide the bulk of the water resources for many major and several minor land masses.

Vegetation patterns and soils are influenced greatly by water availability, and, with topography and geology, largely determine the quantity and distribution of animal life throughout the globe.

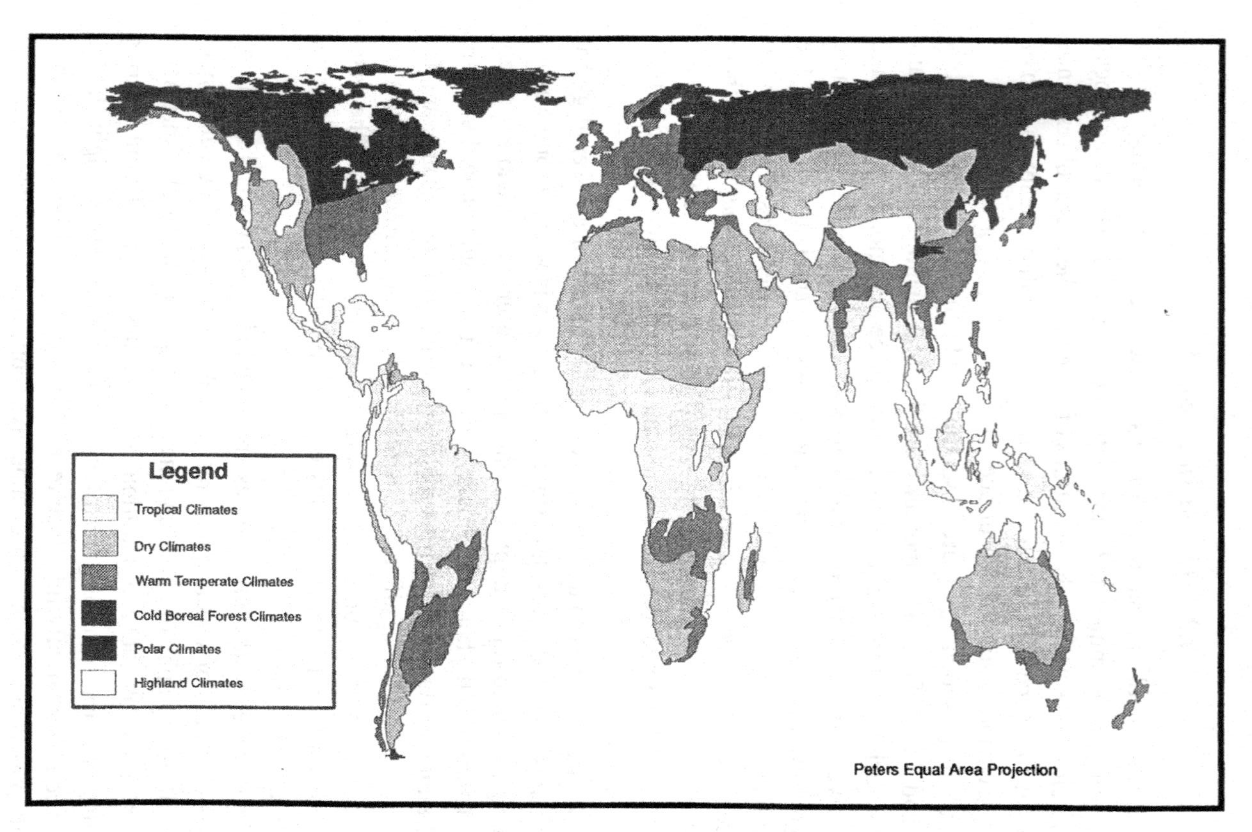

Figure 1. World climate zones (after Strahler and Strahler, 1987, p. 544).

Further modification of how water behaves once it has reached the surface is influenced by geology. The characteristics of the rocks define how quickly water will pass over or through the surface and will largely define the extent and characteristics of underground water storage. The chemical and physical character of the rocks will also help to determine soil character and resulting vegetation patterns.

The quantity, distribution and availability of resources other than water are significant in setting the global scene (World Resources Institute, 1990). Forest resources, whether tropical hardwoods or temperate softwoods, are critical to many national economies. The current debate on the depletion of this resource with ramifications both for local development and for the production of greenhouse gasses and thus climate change has highlighted the importance of this particular resource. Mineral resources are not evenly distributed worldwide, but tend rather to be locally concentrated. The fossil fuels of oil and coal are particularly significant as the basis of much of the world's energy production. Ownership of resources and rights to the exploitation of resources are clearly of fundamental importance to the distribution of wealth and power and underlie development possibilities. In this regard water is a resource of fundamental importance.

1.2.2 The human setting

There are very many aspects to the human condition and the treatment here must of necessity be brief and focused on only a few of the more important aspects.

The characteristics of global population constitute without doubt the most important single factor in the appreciation of global realities, as described in a series of articles in *Ambio* (1992). According to recent estimates (UN-DIESA, 1991(b)) the global population is increasing at a rate which is threatening our existing institutions and has very fundamental implications for the very survival of human-kind. Present distribution, growth projections for 1950–2025, and regions of rapid change are shown in Figures 2 to 4. In large measure, changes are accounted for not so much by changing birth rates (which are in most regions slowly declining) but by increasing longevity, particularly in developing countries. The development and widespread availability of drugs to eradicate many of the diseases which formerly devastated populations have resulted in greatly increased life spans in many countries, which in turn mean rapid increases in overall numbers. The rate of population growth is expected to be far greater in many developing countries than in the highly industrialized countries. Within most countries, but especially within many developing countries, there is expected to be a far higher growth rate in cities than in the surrounding countryside. Migration from the rural to urban areas will be the main reason for this preferential growth. By the year 2000 it is expected that there will be some 22 'mega-cities' with populations

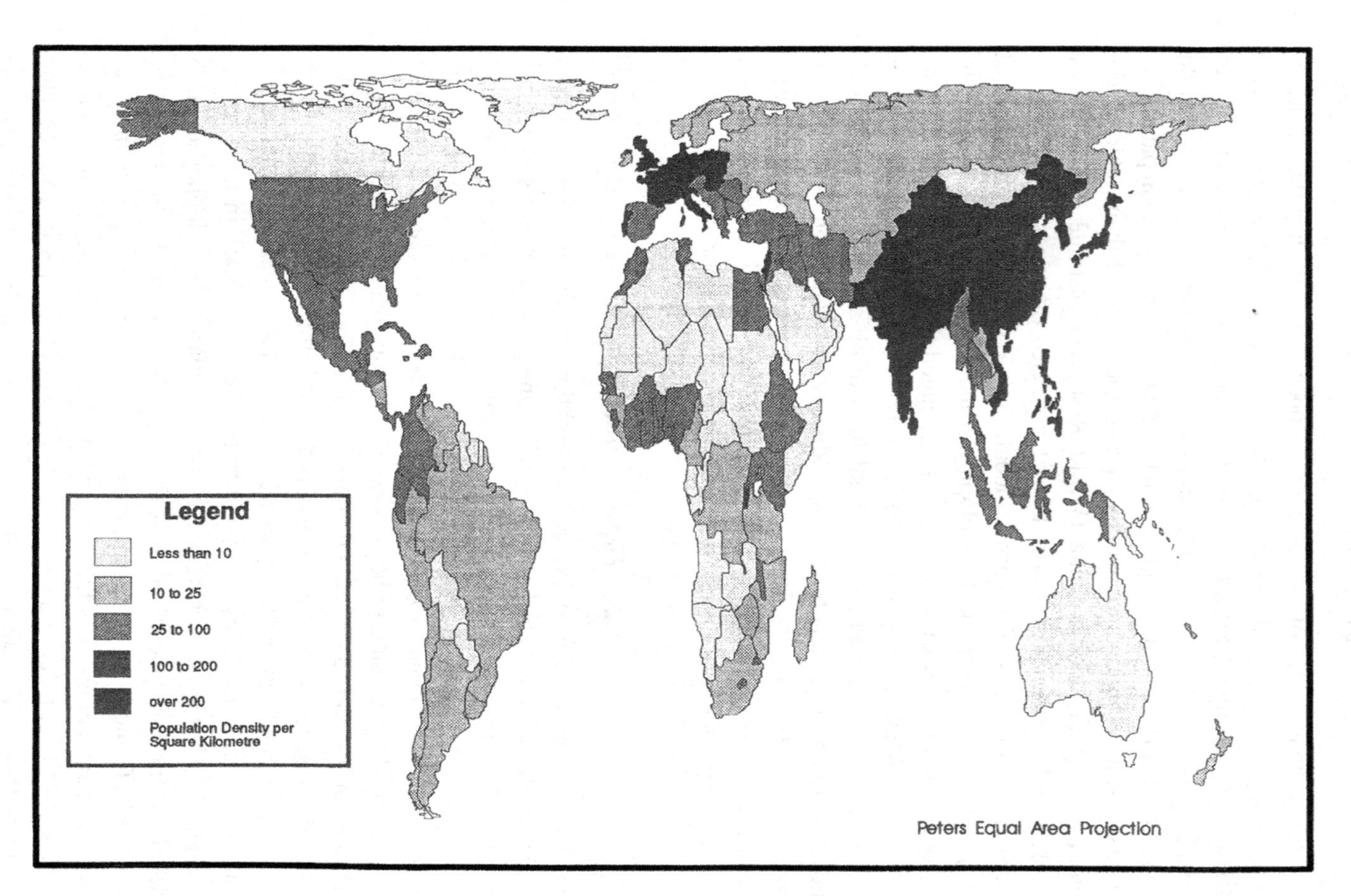

Figure 2. World population density (after Peters, 1990, p. 130).

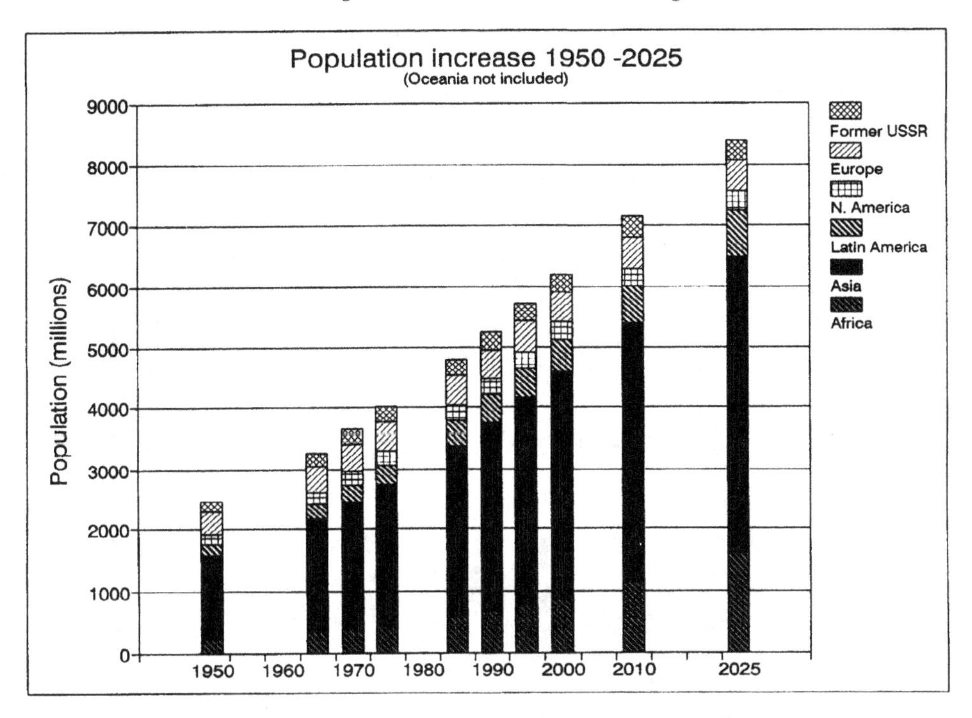

Figure 3. Population growth projections 1950–2025. (Data from World Population Prospects 1990, UN.)

in excess of 10 million inhabitants, as illustrated in Figure 5. Some of these cities will have populations well in excess of 20 million people and 18 of these cities will be in less-developed countries.

The expected doubling of global population within 50 years will have far reaching impacts on natural resources. Demands on the resource base will also increase as societies attain higher standards of living. At present the average individual in western Europe or north America consumes some 30–40 times as much as the average person in the least-developed world. If and when living standards rise in other parts of the world so, too, will demands for natural resources. Clearly, such patterns of consumption cannot be sustained forever. Lifestyles and expectations must change towards a more acceptable distribution of wealth.

Race, religion and political creed are highly variable elements and account for many of the very different ways in which societies conduct their affairs. They are the basis for many of the disagreements and misunderstandings between nations and individuals. Religious beliefs can greatly affect the way in which a society regards its natural resources. Religion, political creed, laws and systems of governance determine the ways open to nations for defining courses of action to solve problems.

The global distribution of wealth and power is highly non-uniform. Disparities between countries are very great indeed. The contrasts in material wealth enjoyed

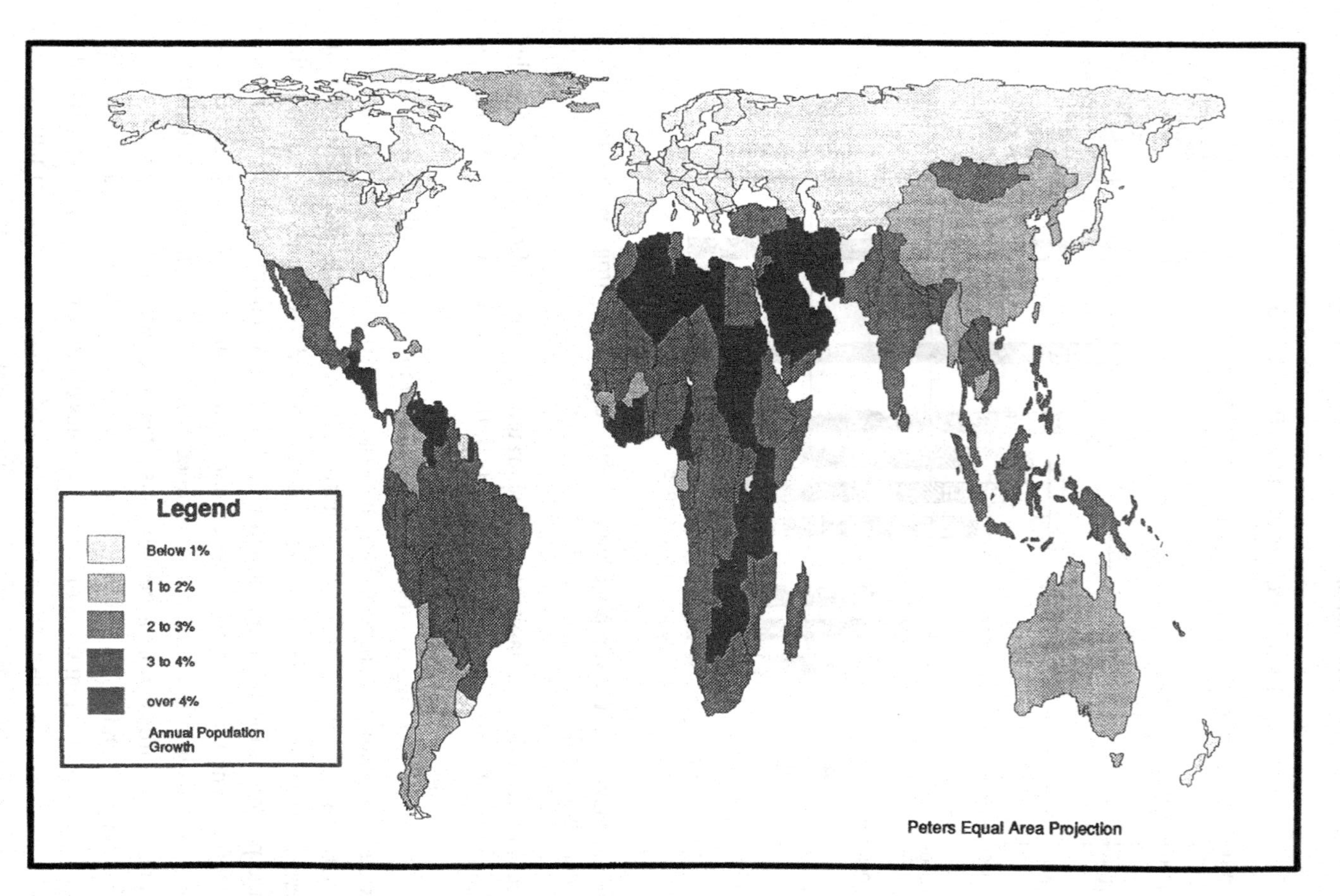

Figure 4. World population change (after Peters, 1990, p. 130).

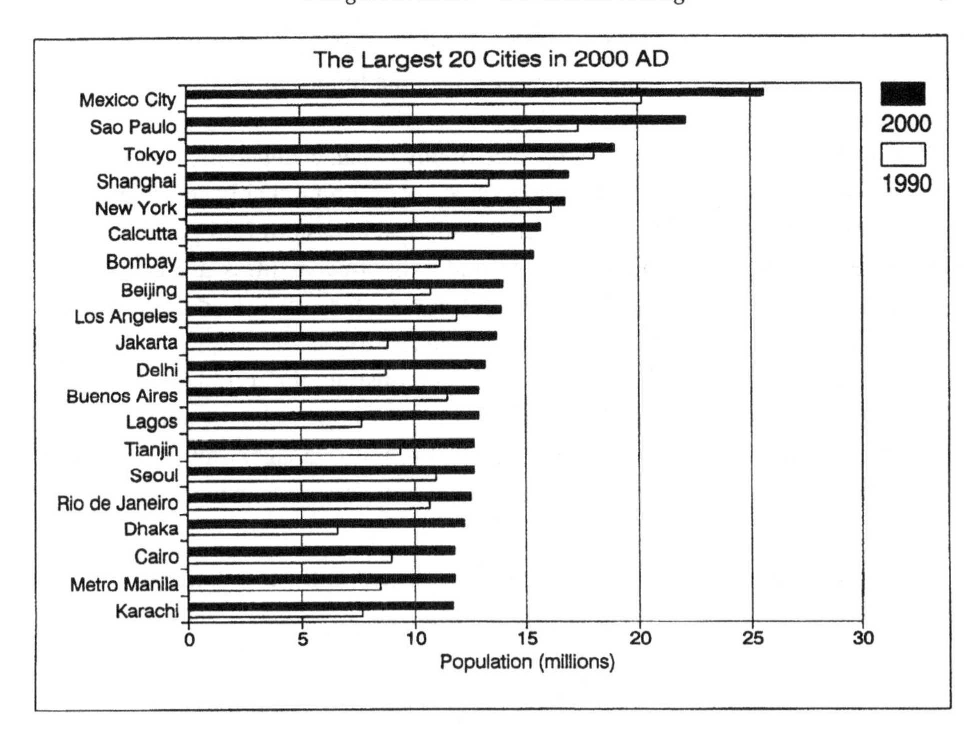

Figure 5. Population of world mega-cities. (After UNCHS-HABITAT/WHO/IBRD/UNDP/UNICEF, 1991.)

by the inhabitants of the developed nations compared with the people in the less-developed countries, is very large. Material wealth and power are closely correlated. The power of large corporations and the control of financial institutions lies almost completely within the more-developed countries. The flow of natural resources is largely from the less-developed to the more-developed countries, thus perpetuating and accentuating these disparities.

There is also disparity in wealth and power within most countries. In all highly developed countries and in the majority of less-developed nations, material wealth and much power is concentrated in the hands of relatively few individuals. This situation is unlikely to change as personal greed and the greed of individual nations, or societies, counteracts any tendency towards a more equitable distribution of wealth and power.

Wars are often the result of disparities in wealth or power. Seemingly an integral part of the history of humankind, wars are currently taking place in several dozen localities on the globe, while more insidious disturbances exist in others which threaten established governments. Clearly, wars are tremendously disruptive, destructive and expensive. Arms production is incredibly costly and the global cost, estimated to be of about US$1.6 million per minute, at the present time, is truly devastating to the betterment of the world's poor. Needless to say, environmental and developmental concerns clearly become totally forgotten in the midst of warfare.

The global military and political balance of power can change very quickly. The collapse of the former Soviet Union took place over a remarkably short period of time. Its repercussions within its former boundaries, in eastern Europe and around the globe, are enormous. New political and economic realities are difficult to assess, reaction to them poses major new thinking for world leaders. Resource management is also affected in a very fundamental way.

This brief overview has been designed to demonstrate that problems besetting environment and water managers and decision makers must be viewed in a very wide-ranging context. Far more than ever in the past, managers must be able to deal with total contexts rather than with isolated problems and small sectors only.

1.3 Water as fundamental to environment and development

1.3.1 Water as the basis for life

Within the biosphere, that thin shell surrounding the globe extending up into the atmosphere, through the oceans and down some few kilometres below the Earth's surface, water is always present. In the atmosphere it is found as vapour, as liquid droplets and as small ice particles. On the Earth's surface it is found as snow, ice and, primarily, in liquid form. In the soil it is also found in all three forms, while in deeper aquifers it is usually only found as a liquid under pressure.

Driven by the energy from the sun, water moves through the biosphere, transporting and redistributing heat and, because many chemical substances can dissolve in water, it transports chemicals too. Almost all life forms depend on water. Life began in the oceans and without water life, as we know it on Planet Earth, would cease to exist. Water may be regarded as a lubricant transporting nutrients from one part of the system of life to another. It has been likened to a liquid mortar, holding all parts of the biosphere together.

Life prospers and declines according to the abundance of water; consequently the different life forms have become adapted to varying amounts of water. Thus, each life form, whether plant or animal, has a developed tolerance for a certain variability in the amount of water available. In times of drought, the less-tolerant species may die out, while in times of flood or super-abundance of water life may also suffer. Species have also come to tolerate a certain mix and concentration of chemicals. If the chemical mix changes, often within the medium of water, and commonly due to human activities, then tolerance limits may once again be exceeded and life will suffer. Thus all ecosystems, or groupings of life forms within a particular part of the overall environment, are vitally dependent on particular quantities of water passing through the system with a certain speed and regularity and maintaining a specific chemical mix. The productivity and diversity of freshwater ecosystems, riverine systems, lakes, wetlands, estuaries and like areas are threatened when the balance of this mix is altered.

As water is so all pervasive and of such importance in so many ways, it is not surprising that in many cultures it is regarded with spiritual reverence. In Hinduism, for example, Varuna is the lord of water and all rivers are manifestations of goodness. In many cultures, water is regarded as indispensable to mental well-being and is often the basis of many tourist resorts. The natural beauty of many rivers and lakes and the inspiration afforded by such spectacular sights as the Victoria or Niagara Falls is a priceless natural heritage. Then there are the numerous spas and baths that provide relief of suffering and cures from taking the waters.

1.3.2 Water as the basis for development

While water is the basis for maintenance of the natural environment, it is also basic to human activities. Thus, it is of critical importance for all socio-economic development.

Basic human needs are all intimately dependent on an adequate supply of water with certain minimum standards of quality. All food production, whether derived from plants or animals (including fish) requires water. The maintenance of basic health standards is linked to adequate supplies of water of sufficient quality. An estimated 80% of all diseases and over one third of deaths in developing countries are caused by the consumption of contaminated water. Proper sanitation to protect against disease requires sufficient water (see the *Special Issue* of *Water International*, Sept., 1991).

Drought conditions, so prevalent in many sub-tropical regions, and of increasing occurrence in many other regions as well, disrupt drinking water supplies and make the maintenance of sanitation systems more difficult as well as affecting agriculture and the food it produces. Falkenmark *et al.*, 1990, make the case that water scarcity is an ultimate constraint in Third World development. Too much water, in the form of floods, can bring sudden death and destruction by destroying dwellings and the various structures which allow people to maintain their economies. It is no wonder that many of the largest man-made structures are dams, designed in large measure to regulate otherwise highly variable flows and so mitigate against the destructive power of floods and the life threatening conditions of drought.

Water is fundamental to virtually every economic activity. Water is withdrawn from natural systems, from rivers and lakes on the surface and from underground aquifers for a variety of uses. Some 80% of the water withdrawn is used for irrigation. Large quantities are also consumed for industrial and municipal use. Much water withdrawn from natural systems is eventually returned, but it may have been contaminated or it may have been thermally changed, making its reuse further down the natural pathway impossible or difficult.

Water is used in large quantities in the generation of electricity. Hydro-power schemes do not consume water, but they have the effect of changing the flow regimes of rivers, which may prove disruptive in other sectors of the economy. Thermal and nuclear power generation also rely on adequate and reliable supplies of water, mostly for cooling, and they are usually located on large rivers and lakes.

1.3.3 The necessity of considering environmental and developmental issues together

In the last few decades there has been an increasing awareness that all issues concerning the environment and development are intimately inter-related, (IUCN/UNEP/WWF, 1991; Kindler, 1992; Falkenmark and Lundqvist, 1992; Koudstaal *et al.*, 1992; Plate, 1992). It follows that all planning and management should be performed in an integrated manner. The water manager has a particularly difficult task as the subject of his or her work is relevant to practically every major human endeavour. But this broadening of the water manager's role and responsibility, while absolutely essential, is putting a great strain on his or her ability to perform adequately. This, at a time when in most economies there are financial restraints on activities and the diminishing numbers of personnel available to do the necessary work, means that performance often lags far behind expectations and requirements.

1.4 The nature of water

1.4.1 The distribution of water in time and space

Water is found in the biosphere in gaseous, liquid and solid forms. It is distributed in the oceans, on the surface of the land, in the rocks beneath the surface and in the atmosphere. The approximate quantities in the various locations are shown in Table 1. The word approximate is used to emphasize that knowledge of the quantities of water that make up the hydrological cycle is poor and inadequate. Most of the freshwater available for human use is found on or very near to the surface of the land. Smaller quantities overall (although locally sometimes of great importance) are available from aquifers under the surface and very small quantities are derived from sea water.

There is a very uneven distribution of water over the land surface, as illustrated in Figures 6 and 7. The arid desert regions contrast with the humid tropics and are different again from the moderate precipitation belts found in mid latitudes. Some

Table 1. *Approximate quantities of water in the various parts of the hydrological cycle with replacement periods*

Category	Total volume ($km^3 \cdot 10^3$)	% of total	% of fresh	Annual volume recycled (km^3)	Replacement period
Oceans	1338000	96.5		505000	2650 y
Groundwater to 2000 m	23400	1.7		16700	1400 y
Predominantly fresh groundwater	10530	0.76	30.1	—	—
Soil moisture	16.5	0.001	0.05	16500	1 y
Glaciers and permanent snow	24064.1	1.74	68.7		—
Antarctica	21600	1.56	61.7	—	—
Greenland	2340	0.17	6.68	2477	9700 y
Arctic Islands	83.5	0.006	0.24	—	—
Other mountain areas	40.6	0.003	0.12	25	1600 y
Ground ice (permafrost)	300	0.022	0.86	30	10000 y
Lakes	176.4	0.013		10376	17 y
Freshwater lakes	91	0.007	0.26	—	—
Salt water lakes	85.4	0.006		—	—
Marshes	11.47	0.0008	0.03	2294	5 y
Rivers	2.12	0.0002	0.006	49400	16 d
Biological water	1.12	0.0001	0.003	—	—
Atmospheric water	12.9	0.001	0.04	600000	8 d
Total water	1385984.61	100[a]			
Total freshwater	35029.21	2.53	100[a]		

[a] Some duplication in categories and sub-categories.
After Kalinen and Bykov (1969), Korzun (1974), L'vovich (1974) and Nace (1969).

regions are characterized by the even distribution of precipitation through time, while other regions display marked seasonality and variability of precipitation over longer periods of time.

Primed by precipitation and greatly affected by temperature, water movement over the land surface is controlled by topography, geology and vegetation cover. Snowfall in cold and temperate regions results in temporary storage of precipitation before its release to streams through melting. Evaporation, especially in hot dry regions, greatly affects runoff. Thus, water moves through the 'hydrological cycle' from the atmosphere to the land surface, over land and underground to the oceans and back to the atmosphere. There are short circuits to this process, temporary storages *en route* and very different residence times for water in the various parts of the cycle. These are depicted in Figure 8 and Table 1.

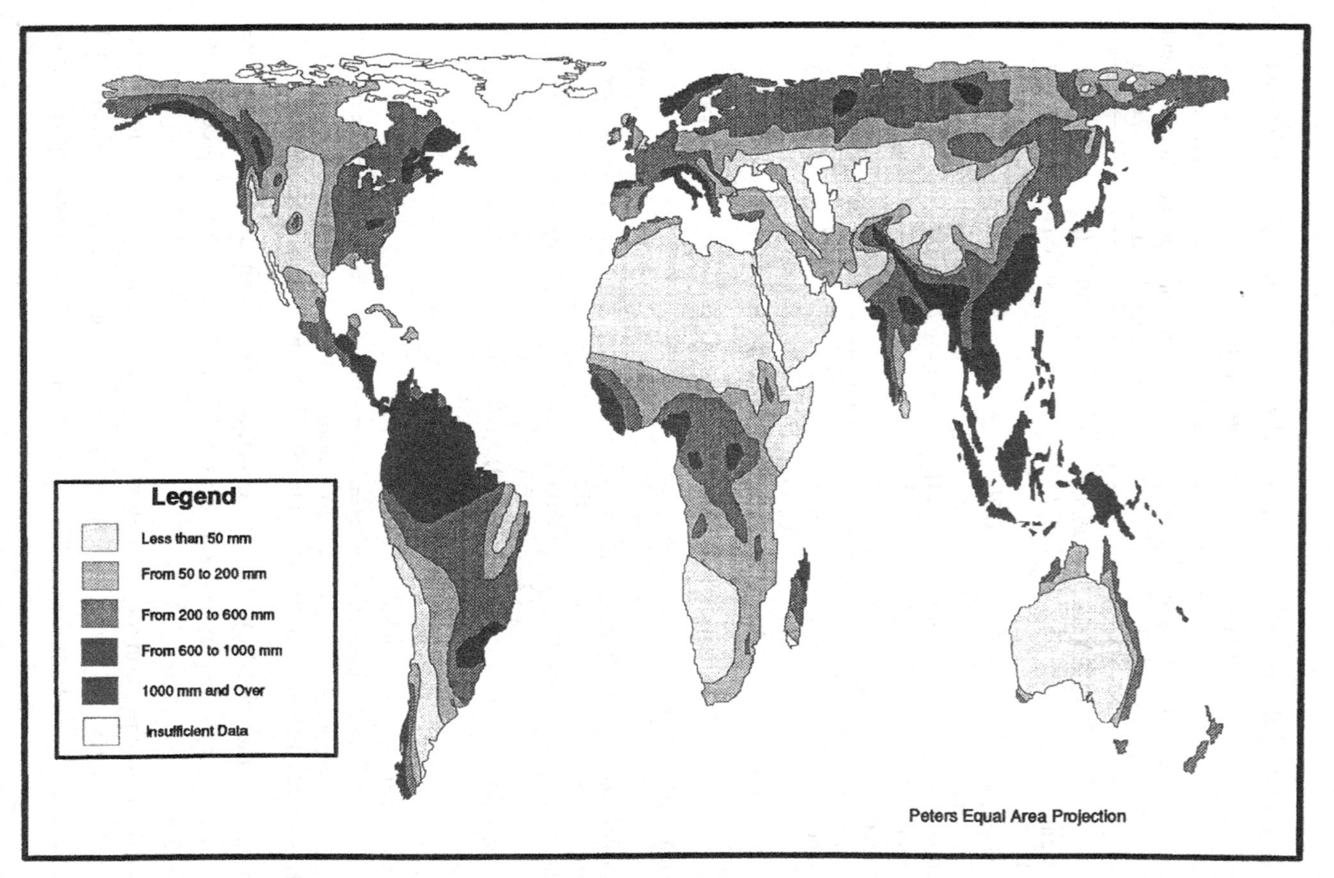

Figure 6. World runoff (after World Resources Institute, 1990, p. 168).

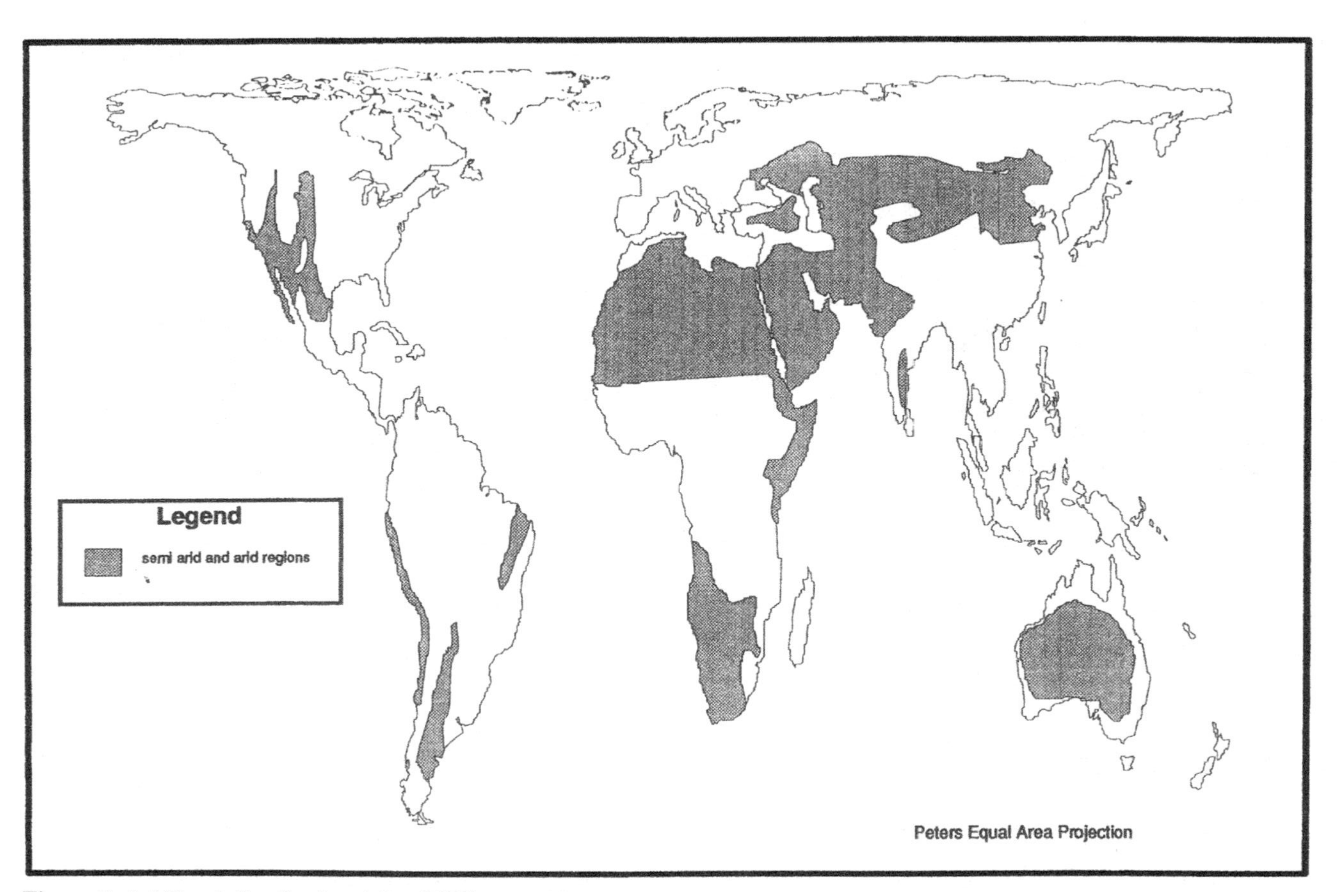

Figure 7. Arid land distribution (after Williams, 1987).

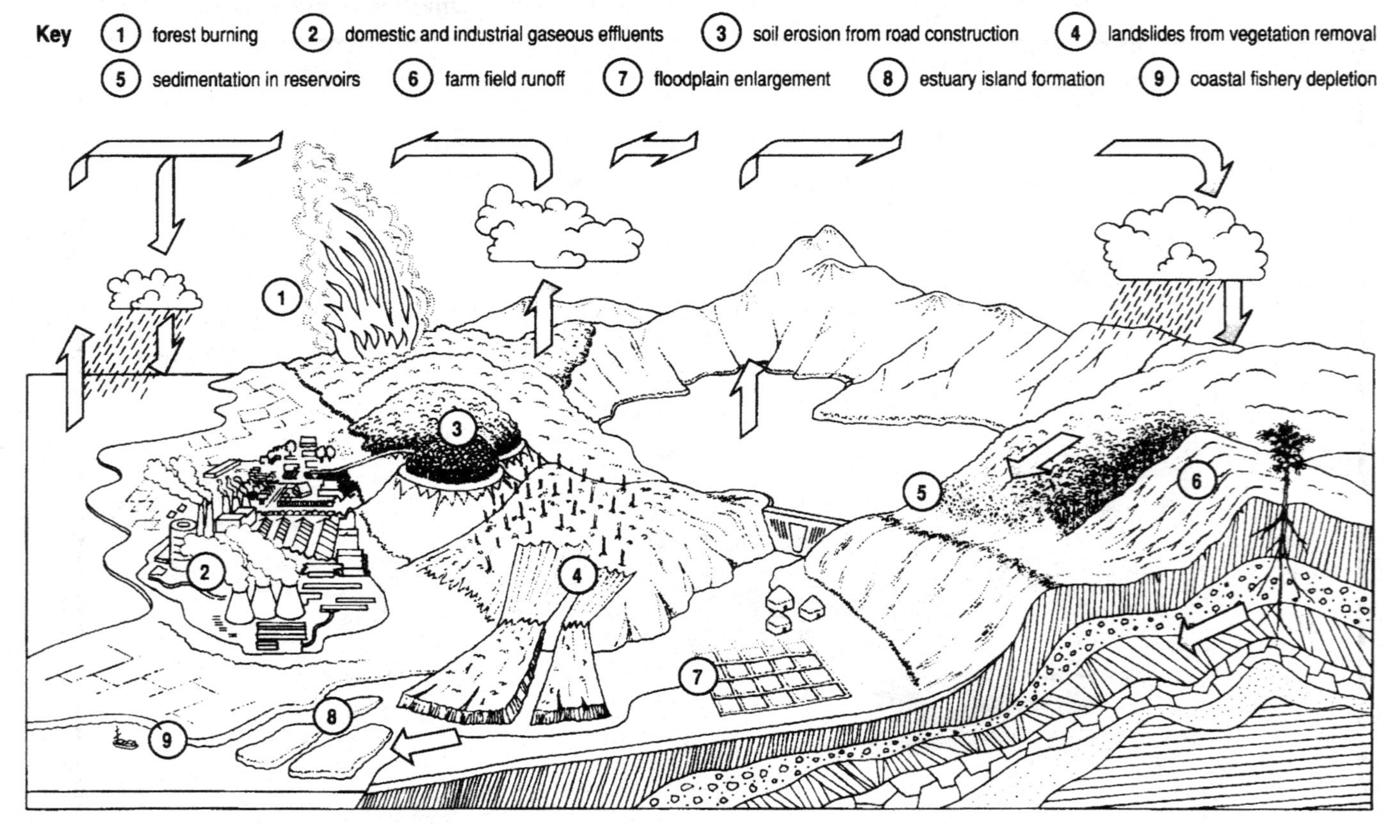

Figure 8. The hydrological cycle. (after WMO/UNESCO, 1991(a), p. 13).

Long-term trends which may significantly alter the hydrological cycle in particular parts of the world are, of course, very important. Thus, the possibility of human-induced climate change with concomitant changes in energy balances and water resources is of great importance. Human activities will be affected by climate change primarily through changes in water availability, but sea level rise will also be important.

Apart from indirect effects on water resources through climate modification, humans influence the hydrological cycle in many direct and often very significant ways. The most obvious effects are by the creation of dams on rivers. Built to regulate flow by creating reservoirs, or to divert water from one river basin to another, such structures augment flow during dry periods and reduce the effects of floods in times of over-abundance. They have other important effects on evaporation and on sediment production and transport, too. Pumping of water from underground aquifers and changes in land use, from the cutting down of forests to the construction of cities, have profound effects on the pathways which water takes and on the rate of its movement. Indeed, virtually all human activity is reflected in changes in the quantity and quality of runoff and groundwater storage in some way or other.

1.4.2 Water as a natural resource

Water is but one of a large number of natural resources, but it is worth considering the characteristics which distinguish water from other natural resources. These characteristics will, in large measure, determine how the resource can be used and they will affect, too, the nature of jurisdiction over the resource and the ways in which its use can be monitored.

Water is one of the few renewable resources. The term 'hydrological cycle' implies renewal but the length of time between the completion of the cycle varies very greatly from one locality to another. Some water moves quickly through the entire cycle while other water may be detained in surface or underground reservoirs for extended periods, but the cycle will eventually come full circle and renewal will take place. In this sense water is somewhat like natural vegetation which will go through cycles of decay or destruction and subsequent renewal. In this it is unlike most mineral resources which, having been mined, are non-renewable within the historical time frame.

Water is mobile. On reaching the Earth from the atmosphere, it moves over and under the surface, some eventually reaching the ocean, some evaporating directly back into the atmosphere, some returning to storage. It moves along restricted and predictable pathways which, under natural conditions, change only very slowly through time. On the surface those pathways converge and form stream networks within natural river basins. The river basin is the fundamental unit within the freshwater world. Water in the headwater reaches will eventually move through the system to exit at the river mouth into sea or ocean. Thus any characteristics which the water has, or attains, within the headwaters will eventually be transmitted to the lower reaches of the basin. Reverse transmission does not take place. Under natural circumstances surface flow does not take place from one basin to another. Transfers may take place underground, but such transfers are usually small and slow.

The fact that water is mobile implies that it may move from one political jurisdiction to another. This is so at the very local scale when a stream flows from one person's property to another; at intermediate scales when the flow is from one local or provincial jurisdiction to another; and at international scales from one country to another. As the use or misuse of the water in the headwaters affects the quantity and quality in the lower reaches, the characteristic of mobility is at the root of many of the problems concerning jurisdiction over the resource.

In liquid and solid phases the mobility of water is confined within river basins. This is of fundamental importance. What happens to water within one river basin usually has no or little direct bearing on what happens to water within another basin unless there are inter-basin transfers. Thus, the management of water on one continent has no direct bearing on the management of water on another continent. In this respect water is unlike air, which is also highly mobile. A polluting gas discharged into the atmosphere in one locality mixes relatively quickly to affect wide regions and soon the entire globe. A pollutant entering a river will contaminate surface runoff solely in that part of the basin through which the polluted water will move. Contiguous basins will only be contaminated through groundwater movement or by inter-basin transfers. Receiving seas and oceans will, of course, receive pollutants derived both from overland and groundwater flows. Atmospheric transport of pollutants and their wet and dry deposition can, however, affect water quality over wide areas. The phenomena of acid rain and related pollution transfer mechanisms are well known, most especially in the industrialized world.

Mineral resources are static. They are found in particular localities or regions. Their ownership and the jurisdiction over their exploitation is usually far less complex than the ownership and jurisdiction over water. Natural vegetation resources are found in fairly distinct areas or regions. Soils, closely linked to natural vegetation are also found within particular zones. Over long periods of time the boundaries of such regions may migrate, but with a few notable exceptions, such as the rapid encroachment of desert in some areas and human-induced changes in the boundaries of forest areas, vegetation zones are relatively static and response to climatic changes is typically slow. Wildlife resources are mobile. Animals are mobile within natural ranges which may be very restricted or widespread but are usually within continental confines. Birds and sea-life can be very wide-ranging.

The concept of ownership or rights over wildlife and vegetation is, of course, differently interpreted in different regions of the world. However, in general, ownership and jurisdiction over wildlife is closely affected by its inherent mobility.

The simple but profoundly important distinctions between the characteristics of natural resources such as have been outlined are rarely given sufficient emphasis in discussion of water as a natural resource. Their importance in understanding questions of ownership of and jurisdiction over natural resources cannot be overstressed.

Water is used in a great variety of ways. Demand for the resource in many parts of the world is outstripping supply. A summary of major use categories and an indication of changes in demand over time on a global basis is given in Table 2.

1.4.3 The river basin as a natural unit

As mentioned in 1.4.2, water on the surface of the Earth is naturally organized within river basins. Underground aquifers may or may not coincide with surface basin delineations. Many hydrologically important aquifers are, however, shallow and interact with surface water movement wholly or largely within the confines of the surface basins that they floor.

The significance of the river basin as a natural unit is very important, as water interacts with and to a large degree controls the extent of other natural components in the landscape such as soils, vegetation and wildlife. Human activities, too, being so dependent on water availability, might best be organized and coordinated within the river basin unit.

River basins exhibit great differences in size and shape, ranging from the very small catchments draining directly into the oceans to the largest catchments such as the Nile (3.03×10^6 km^2) or the Amazon (5.87×10^6 km^2). Basins are, of course, found within all climatic zones with the result that total discharges and the regime of flows from season to season are quite different between basins.

It makes sense that the management of natural resources and the management of those human activities which are largely dependent on water should be undertaken within river basin units, or be sub-divided according to sub-basin delineations within major catchments. However, as illustrated in Table 3, the reality within many river basins is that political boundaries do not coincide with the confines of river basins. While this fact is important on all scales of consideration from the very local through regional to international, it is of greatest importance in international transboundary situations. There are more than 200 major river basins worldwide which fall within more than one country, many occurring within several countries. The numbers grow with political fragmentation, such as that recently occurring within the former Soviet Union.

Table 2. *Approximate world water demand ($km^3\ y^{-1}$) according to use*

Water users	1900	1940	1950	1960	1970	1980	1990	2000
Irrigated area (Mha)	47.3	75.8	101	142	173	217	272	347
Agriculture								
A	525	893	1130	1550	1850	2290	2680 (68.9)	3250 (62.6)
B	409	679	859	1180	1400	1730	2050 (88.7)	2500 (86.2)
Industry								
A	37.2	124	178	330	540	710	973 (21.4)	1280 (24.7)
B	3.5	9.7	14.5	24.9	38.0	61.9	88.5 (3.1)	117 (4.0)
Municipal supply								
A	16.1	36.3	52.0	82.0	130	200	300 (6.1)	441 (8.5)
B	4.0	9.0	14	20.3	29.2	41.1	52.4 (2.1)	64.5 (2.2)
Reservoirs								
A	0.3	3.7	6.5	23.0	66.0	120	170 (3.6)	220 (4.2)
B	0.3	3.7	6.5	23.0	66.0	120	170 (6.1)	220 (7.6)
Total								
A	579	1060	1360	1990	2590	3320	4130 (100)	5190 (100)
B	417	701	894	1250	1540	1950	2360 (100)	2900 (100)

A: Total water consumption, B: Irretrievable water losses. Percentage figures are given in parentheses.
From Shiklomanov (1991).

Island states such as Australia or Sri Lanka do not have to be concerned with problems of transboundary flows between nations, which are of importance in such countries as those sharing the waters of the Rhine or the Danube. But these problems are of far greater importance in many of the basins of major rivers in Africa, the Middle East and in many Asian countries, in which water may be limited in supply and where demands on the resource are great and will be increasing. In such basins as those of the Nile and the Tigris–Euphrates the scarcity of water and the splitting of its ownership and jurisdiction between several countries has been and likely will be the basis for international conflict. 'Water Wars' are, unfortunately, likely to be of more and more common occurrence in the future.

Within basins embracing several nation states, the attitudes of those states in the headwaters is typically different from the attitudes of the states in the lower reaches of the river. Headwaters states do not wish to be held to guarantees of water quantity or of water quality demanded by their downstream neighbours. Downstream countries are, very naturally, concerned with the prospects of any

Table 3. *Selected river basins with the percentage of the area within each country*

Amazon (5 870 000 km^2)		Mekong (802 900 km^2)		Euphrates/Tigris (778 834 km^2)	
Brazil	63.3	Lao PDR	25.4	Iraq	59.1
Peru	15.9	Thailand	22.9	Turkey	19.6
Bolivia	11.9	China	22.2	Iran	14.1
Colombia	5.8	Dem. Kampuchea	18.9	Syrian Arab Rep.	7.2
Ecuador	2.1	Vietnam	7.7		
Venezuela	0.9	Myanmar (Burma)	2.9		
Guyana	0.1				
Nile (3 030 700 km^2)		**Niger (12 150 00 km^2)**		**Zaire/Congo (3 457 000 km^2)**	
Sudan	62.7	Mali	28.2	Zaire	62.1
Ethiopia	12.1	Nigeria	26.4	Central Af. Rep.	10.9
Egypt	9.9	Niger	22.3	Angola	7.2
Uganda	7.7	Algeria	6.8	Congo	6.9
Un. Rep. Tanzania	3.8	Guinea	4.3	Zambia	4.7
Kenya	1.8	Cameroon	4.1	Un. Rep. Tanzania	4.5
Zaire	0.8	Burkina Faso	3.6	Cameroon	2.7
Rwanda	0.7	Benin	2.3	Rwanda	0.1
Burundi	0.5	Cote d'Ivoire	1.1	Burundi	0.4
		Chad	0.9		
Zambezi (14 19 960 km^2)		**Danube (81 5850 km^2)**		**Rhine (185 000 km^2)**	
Zambia	40.7	Romania	29.3	Germany	54.5
Angola	18.3	Yugoslavia (former)	22.5	Switzerland	15.3
Zimbabwe	15.9	Hungary	11.7	Netherlands	13.5
Mozambique	11.4	Austria	10.0	France	12.8
Malawi	7.7	Czech and Slovak Republics	8.2	Austria	1.6
Botswana	2.8	Germany	7.0	Luxembourg	1.4
Un. Rep. Tanzania	2.0	Bulgaria	5.3	Belgium	0.8
Namibia	1.2	USSR (former)	5.2	Liechtenstein	0.5
		Switzerland	0.4		
		Italy	0.3		
		Poland	0.05		
		Albania	0.02		

upstream dam construction which may affect the regime or total quantity of flow, especially by major diversions of water into or out of the basin. They are concerned, too, with any upstream water withdrawals or any polluting of the waters which will affect downstream usage. Figure 9 illustrates some of the major 'multi-national' basins in which there are significant transboundary jurisdiction problems.

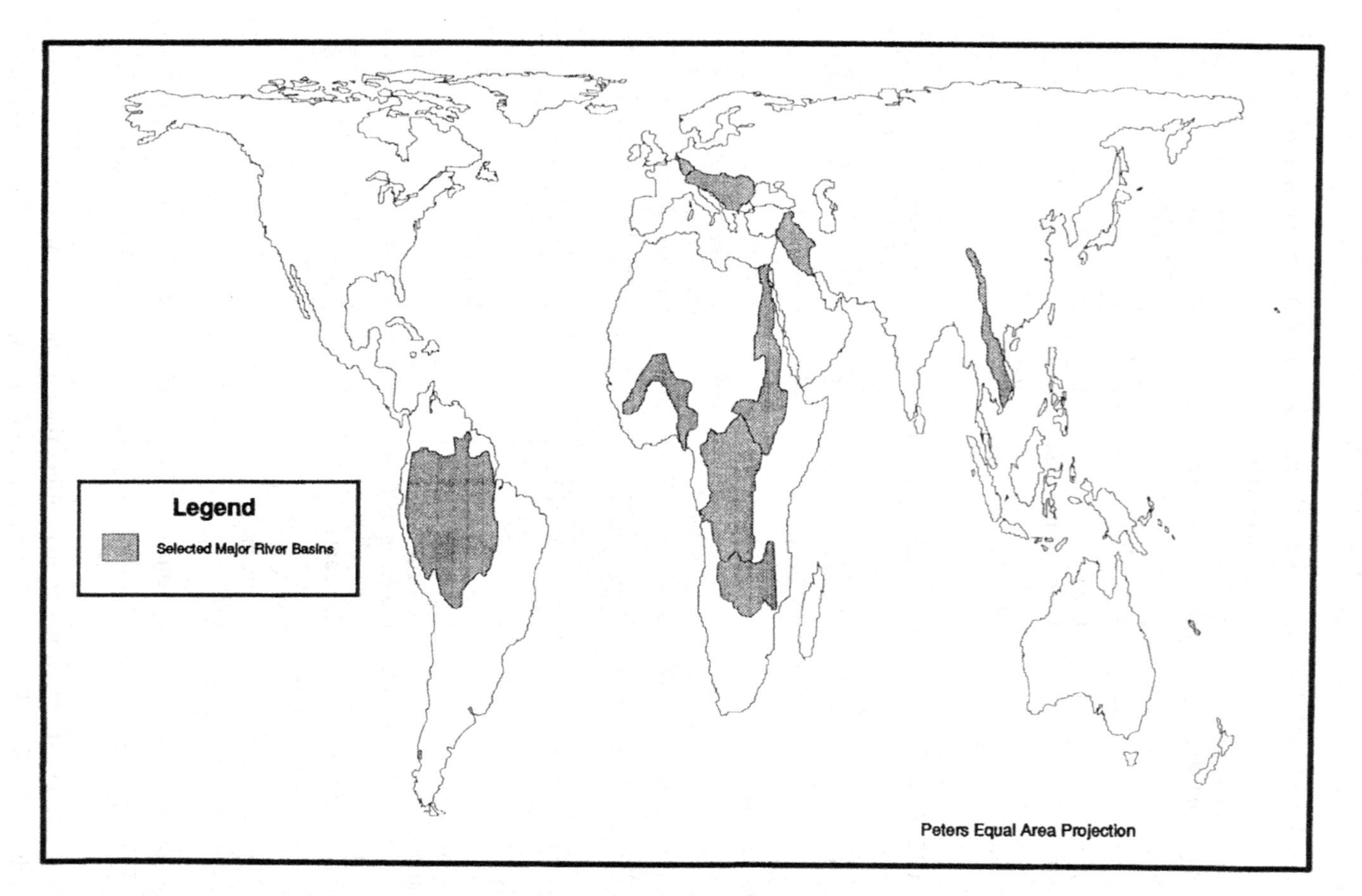

Figure 9. A selection of major international rivers (after Snead, 1972, p. 64).

1.5 Differing perspectives

Everyone is concerned with water and thus everyone has a stake in how the resource is managed. Individuals and groupings of people have very different perspectives on water problems and on the best ways of management depending on the viewpoint and the bias of the group. Management is usually carried out within political frameworks, whether on the local, the provincial, the national or the international levels. Within and between these political entities other groupings come into play: bilateral and multilateral agencies linking countries together; United Nations agencies and bodies usually focusing on particular sectors within the water domain; non-governmental organizations promoting a particular viewpoint or group of people; religious and other advocacy groups promoting specialized points of view.

The most important single entity in water resources management is the nation state. In most countries overall responsibility for water management rests at the national level. There is commonly a decentralization of responsibility to provincial and local levels. Because so many decisions on water management must, to be effective, be made at the local or provincial level, considerable responsibility often lies at these levels. However, overall coordination and any negotiation or interaction internationally is handled through national institutions of government.

Nations interact with each other through these institutions and through a number of other mechanisms. Most developed countries have international departments or agencies through which government aid and technological assistance in the water sector are channelled. Such assistance can be of great significance and of considerable quantity and, normally as a consequence of historical relationships, is channelled in particular directions.

United Nations agencies and bodies dealing with water are many and varied. There is no single agency dealing with water. Rather, there are a number of agencies each dealing with a particular aspect of water and some regional UN bodies which coordinate concerns of water within a broader regional context. A listing of UN agencies with water interests is found in Table 4.

Countries cooperate within a number of other groupings. Countries with like circumstances or problems come together within such organizations as the Organisation for Economic Co-operation and Development (OECD), which links the most developed countries together. Especially important in resolving regional problems are river basin commissions such as those of the Rhine, the Mekong and the Niger. In fact there are a large number of river basin commissions worldwide with a variety of shades of power for controlling their basins, agreeing basin-wide policies and settling disputes, but, unfortunately, not one commission for each and every international river basin.

Table 4. *Members of the United Nations Administrative Committee on Coordination Inter-Secretariat Group for Water Resources (ACC/ISGWR) listed in UN order of precedence (December 1991)*

UN Department of International Economic and Social Affairs (UN-DIESA)
UN Department of Technical Co-operation for Development (UN-DTCD)
UN Children's Fund (UNICEF)
UN Development Programme (UNDP)
UN Environment Programme (UNEP)
UN University (UNU)
UN Economic Commission for Africa (UN-ECA)
UN Economic Commission for Europe (UN-ECE)
UN Economic Commission for Latin America and the Caribbean (UN-ECLAC)
UN Economic and Social Commission for Asia and the Pacific (UN-ESCAP)
UN Economic and Social Commission for Western Asia (UN-ESCWA)
UN Centre for Human Settlement (UNCHS HABITAT)
UN Disaster Relief Co-ordinator, Office of the (UNDRO)
International Research and Training Institute for the Advancement of Women (INSTRAW)
World Food Programme (WFP)
International Labour Organization (ILO)
Food and Agriculture Organization of the UN (FAO)
UN Educational, Scientific and Cultural Organization (UNESCO)
World Health Organization (WHO)
World Bank (IBRD)
World Meteorological Organization (WMO)
UN Industrial Development Organization (UNIDO)
International Atomic Energy Agency (IAEA)
UN Conference on Environment and Development (UNCED)

Other UN agencies and affiliated bodies with interests in water
UN Non-Governmental Liaison Service (UNNGLS)
Organization of African Unity (OAU)
PROWWESS, Afrique/UN Development Fund for Women (UNIFEM)
International Decade for Natural Disaster Reduction (IDNDR)
UN Institute for Training and Research (UNITAR)
UN Research Institute for Social Development (UNRISD)
UN Information Centre (UNIC)
UN Sudano-Sahelian Office (UNSO)
Water Supply and Sanitation Collaborative Council (WSS)

Non-governmental organizations (NGOs) come in all shapes and sizes and with a variety of roles. They range from the very small and local NGOs to the major scientific NGOs found within the International Council of Scientific Unions. Their mandates, their modes of operation and their effectiveness are extremely varied.

There are a number of groups of individuals and institutions brought together under fairly informal organizations, but which are sometimes of notable influence.

Such a group is the Collaborative Council for Drinking Water Supply and Sanitation, which is continuing the work started in the International Drinking Water Supply and Sanitation Decade (IDWSSD).

Minority, ethnic, religious and other groupings, such as the now very common women's groupings, promote specialized positions. The advocacy from these groups can be extremely powerful and more and more the voices from these groups are being heard and are being acted upon. Important among them are womens' groups at local, national and global levels. Women play an extremely active role in many countries in both the very local activities of collecting water and in the overall management of the resource.

Humankind has devised an incredible array of organizations and agencies to deal with water issues. These organizations not only have very different mandates and perspectives but they are made up of individuals who also have their own very different personalities and views. It is no wonder, then, that agreements on which issues should have priority and agreements on solutions to problems are so elusive.

1.6 Institutions and the development of human resources

Given that an integrated approach to water resources management is required in order to properly address a very complex and inter-related set of issues, it follows that the institutional structures within countries should be set up accordingly. But all too often we find that institutional arrangements are far from adequate for the tasks to be performed.

In many countries there are institutions at national, provincial and local levels, often with overlapping responsibilities. In addition, because water is of concern to so many economic activities, many government departments have sections dealing with water. This means that in most countries there is very serious institutional fragmentation, to the detriment of the integrated approach. It seems to be part of the human psyche that ambitious individuals wish to build empires of power and authority around themselves and thus fragmented institutional arrangements are self-perpetuating. It is an unusual situation indeed to find these disparate empires working in complete harmony with each other.

In many countries, to address new 'environmental' thinking and philosophy, new departments of the environment have been created. Central water agencies have not always been incorporated into these new departments to the disadvantage of the integrated management approach. Thus, unfortunately, in many countries the desire to integrate planning and management by the creation of new departments of the environment has had exactly the opposite effect and has resulted in increased fragmentation.

The need for coordinated management argues for a centralization of authority and power. While there are some very strong arguments that this would be efficient, there are equally strong dangers. The effects of most water-related projects, whether they concern pumping from wells, use of water for irrigation or damming a river for hydro power, are felt at the very local level. It follows that many decisions on water projects should be made at the local level or with significant input from local water interests. This question of the most appropriate level of decision making on water-related projects is often very vexing and difficult for water managers and politicians alike to solve.

The theme of the relative weight of individual against collective incentives is also very important. The individual person is always likely to argue in his or her own self interest when it comes to choices promoting one use of water over another. Self interest versus the overall benefits to a larger group within the society is a common theme often being played out. This same theme may be applied at a number of scales. Self interest of one community against another, of one province against another or of one country against another is the basis of argument, disagreement and, in some cases, conflict between parties. One of the fundamental arguments at UNCED was of this type. In a world of very unequal distribution of wealth, resources and power, the self interests of individual countries are often promoted at the expense of the overall good of society as a whole.

There are not only the needs for restructuring in order to promote an integrated management approach, but there are also the needs for blending central coordination with grass roots level management.

At the heart of all organizations are people. No matter how much structural change within organizations is effected, the basic thinking and philosophy of individual workers, and most particularly that of individual senior managers, must change too. A critical question is whether individuals and institutions can change quickly enough so that the new methods of management can be effectively introduced.

1.7 Changes in policy during the last few decades

Perspectives on the environment and on the links which the environment has with development have been changing through time. In the last two decades there has been a rapid growth in environmental awareness which has, not coincidentally, been simultaneous with rapid growth in the efficiency and pervasiveness of telecommunications whereby much of the world's population can be quickly informed of developments around the globe. Through fast exposure to news and opinion from different parts of the world, individuals are better able to realize the differences in circumstances from place to place and understand the complexities of realities facing them as global villagers.

The development in environmental awareness has been greatly affected by a few key events which have focused world attention. Three major events, and countless minor ones, have been of particular importance. The 1972 UN Conference on the Human Environment in Stockholm led to the establishment of the United Nations Environment Programme and to the setting up of Ministries of the Environment in a large number of countries and to the boosting of environmental awareness. Recognition of the fact that environment and economic activities were inextricably linked grew over the following years, but really crystallized in 1987 with the publication of the report of the World Commission on Environment and Development *Our Common Future*. This linkage was emphasized in the four preparatory conferences leading to the UN Conference on Environment and Development held in Rio de Janeiro in 1992. These events have been extremely important in focusing world attention on environmental issues and on raising the levels of awareness worldwide. Major conferences, such as those at Stockholm and Rio, may not achieve everything which was hoped of them, but they do demand attention and they are a means of crystallizing opinion, which is a vital prerequisite to action.

Perspectives on environmental matters vary enormously. Views from the developed countries contrast with those from the developing countries. The developed countries with, by definition, high standards of living for their inhabitants, with well-developed economies and with relatively stable populations, are rich enough to be able to spend the necessary funds on cleansing and maintaining their environments without unacceptable decreases in living standards. Measures to restore environments to acceptable standards are often costly and can take enormous efforts, but if the will of the people and politicians is right, then change for the better can take place. There is a slow but sure change in philosophy (albeit with setbacks in various places) in which humankind is more and more being perceived as part of an ecological whole and in which humans must live in harmony with nature, if *homo sapiens* is to survive for many more generations.

Perspectives from developing countries are very different. Clearly, with such a variety of circumstances and conditions in these countries, it is somewhat dangerous to generalize, but there are views which have a certain commonality. Standards of living are generally very low, disease and malnutrition are common. Resources are generally inadequate and cannot be supplemented from industrial output. Food is often critically scarce. Populations are growing out of control and cities are exploding into their hinterlands fed by large numbers of rural poor forced to flee their agricultural lands because of degradation of their soils, or for economic or other reasons. In these circumstances it is no wonder that perspectives are different.

It is somewhat ironic that many societies, in particular indigenous societies in developing countries, have had a long-standing tradition of viewing themselves as part of nature and of realizing that sustainability in development requires caring for the land on which they live. They adjusted their lifestyles accordingly. Poverty and desperation force many of these people to mistreat their land, taking whatever is necessary for daily survival without being able to save for the future. So called civilization has swept away many of the concepts and ethics on which these societies are based and has nullified the practical application of their philosophies.

As a result of the very pressing needs to address critical development requirements, while being aware of the long-term need to be more environmentally conscious, Third World scientists are developing philosophies in which protection of the productivity of environments is stressed, instead of protection of the environment itself in the more traditional Western sense.

There have been a number of themes running through the evolution of environmental thinking in the last two decades in the more developed countries. First, within the natural environment, there has been a realization that water should not be considered apart from other elements in natural systems. Water interacts with the land primarily through soil and vegetation systems. This land/water interaction is so strong that the one should never be considered without the other (Falkenmark and Lundqvist, 1992). Indeed, it is argued by Falkenmark that, within the concept of the hydrological cycle, the importance of vegetation has not been sufficiently stressed. Perhaps, compared with the horizontal movement of water over the land surface, more stress should be put on the importance of vertical fluxes of water through the biomass cover. This view of far more interrelation between the elements of the natural environment has resulted in an appeal for more ecologically sound management which would result in the better preservation of natural ecosystems.

Second, there has been a change from the view that environmental concern is necessarily contrary to the promotion of development to a view which does not see development being hampered or delayed by consideration of environmental concerns. There is a growing body of opinion that, especially if a long-term perspective is taken, development can only truly proceed if combined with care of the environment.

Third, there has been growth and acceptance that development must be sustainable. The World Commission on Environment and Development in *Our Common Future* defined sustainable development as 'development that meets the needs of the present without compromising the ability of future generations to meet their own needs'. In *Caring for the Earth* (IUCN/UNEP/WWF, 1991) *sustainable development* is used to mean 'improving the quality of human life while living within the carrying capacity of supporting ecosystems'; a *sustainable economy* is the product of sustainable development – it maintains its natural resource base and it can continue to develop by adapting and through improvements in knowledge, organization, technical efficiency and wisdom. A *sustainable society* would live by the following nine principles*:

(a) Respecting and caring for the community of life;
(b) Improving the quality of human life;
(c) Conserving the Earth's vitality and diversity;
(d) Minimizing the depletion of non-renewable resources;
(e) Keeping within the Earth's carrying capacity;
(f) Changing personal attitudes and practices appropriately;
(g) Enabling communities to care for their own environments;
(h) Providing a national framework for integrating development and conservation;
(i) Creating a global alliance.

Adoption of these principles would involve radical changes in the way we conduct our affairs and, in particular, would imply that a truly integrated approach to the use and development of natural resources must take place. These new approaches are described in many publications including ECE (1991) and ESCAP (1991). The implication is not only that within countries and regions integrated management must become the norm, but also that there must be a far greater degree of integration of management between nations. There is also an implication that principles must be developed for encouraging and enforcing more efficient and sustainable modes of action.

1.8 The lead-up to ICWE and UNCED

Between the 1972 Stockholm Conference on the Human Environment and the 1992 Rio Conference on Environment and Development there were a series of conferences and meetings and the preparation of numerous reports in the various sectors of the water field in support of programmes and activities in these fields. They are briefly reviewed here.

Highly important initiatives have been undertaken by the many UN agencies involved in water. These initiatives are very wide-ranging, including the assessment and management of the resource; the development of the resources for agriculture, industry and commerce; the protection of the resource for water supply and sanitation as well as for the protection of the natural environment; and the development of the human potential to make best use of the resource. Some of these have been undertaken on a global basis and others, the majority, have been in the form of technical assistance projects of a regional or national type.

The UNESCO International Hydrological Decade 1965–74 (IHD), succeeded by the International Hydrological Programme (IHP), the Operational Hydrology

* Wording adapted from *Caring for the Earth*.

Table 5. *Topics discussed at ICWE and Mar del Plata*

	Major topics addressed in Working Groups at ICWE, Dublin, 1992	Chapter reference within this book	Major topics addressed at UN Water Conference, Mar del Plata, 1977
(a)	Integrated Water Resources Development and Management	Chapter 2	Policy, planning and management (d)
(b)	Water Resources Assessment and Impacts of Climate Change on Water Resources	Chapters 3 and 5	Assessment of water resources (a) Natural hazards (e)
(c)	Protection of Water Resources, Water Quality and Aquatic Ecosystems	Chapter 4	Environment, health and pollution control (c)
(d)	Water and Sustainable Urban Development and Drinking Water Supply and Sanitation in the Urban Context	Chapters 6 and 8	Water use and efficiency (b)
(e)	Water for Sustainable Food Production and Rural Development and Drinking Water Supply and Sanitation in the Rural Context	Chapters 7 and 8	Water use and efficiency (b)
(f)	Mechanisms for Implementation and Coordination at Global, Regional, National and Local Levels	Chapters 9 and 10	Public information, education, training and research (f) Regional cooperation (g) International cooperation (h)

Programme (OHP) of WMO, the initiatives in environmental health of WHO, the initiatives in food and agriculture of FAO and the numerous technical assistance initiatives of the United Nations Development Programme (UNDP) and the World Bank exemplify the great efforts made in the water sector within UN agencies.

These and other programmes in water led to the 1977 UN Water Conference at Mar del Plata, Argentina. This was the first global water conference involving delegations from governments, non-governmental organizations and the UN family which addressed a broad range of topics. No other similar conference was held until the ICWE in Dublin in January 1992. The Mar del Plata Action Plan stimulated a number of activities including the International Drinking Water Supply and Sanitation Decade (IDWSSD) 1981–90. The UN Inter-Secretariat Group for Water Resources was more formally constituted as a result of the UN Water Conference. The topics discussed at Mar del Plata and Dublin were similar, as shown in Table 5.

Significant studies and programmes have also been undertaken by non-governmental organizations and other bodies. Some of the more important initiatives have been promoted and executed by the International Association of Hydrological Sciences (within the family of the International Council of Scientific Unions) and by the International Water Resources Association. The Collaborative Council for Water Supply and Sanitation, a loosely structured but highly effective forum for discussion of water supply and sanitation (WSS) issues, has developed from the International Decade for Drinking Water Supply and Sanitation, and is continuing to undertake very important initiatives in this area.

Many projects and programmes are coordinated and undertaken within specific regions. Many UN agencies have regional and national offices and, in collaboration with, among other organizations, the regional development banks, projects are implemented within groupings of countries having similar conditions and problems. There are many external support agencies (ESAs) involving single or small groupings of donor countries interacting with individual or small groupings of recipient countries which also act on a regional basis.

1.9 On the organization of the International Conference on Water and the Environment

1.9.1 Concept

In the years before the decision to hold UNCED, there were some discussions between UN bodies and agencies with leading roles in the water field on the need to convene, at some time in the future, a follow up to Mar del Plata. Several more specialized conferences were planned and convened, such as the New Delhi Consultation at the end of the International Drinking Water Supply and Sanitation

Decade*. However, when the decision was made to mark the 20th anniversary of the Stockholm Conference (UNGA Resolution 43/196, 1988) and when 'protection of quality and supply of freshwater resources' featured second in the list of nine priority issues identified in para. 126 of UNGA Resolution 44/228 a year later, the need for a specialized water conference became very evident. The idea was that this would be a conference where experts and professionals well versed in the different fields of water would prepare the major input on freshwater to the UNCED process, with its organization being entrusted to the Inter-Secretariat Group on Water Resources.

The formal decision to incorporate ICWE into the preparations for UNCED was taken during the first PrepComm in Nairobi in August 1990, following the offer of the Government of Ireland to host the meeting. This enhanced the legitimacy of ICWE but in addition it imposed a very tight time schedule on the organizers. This decision also imposed the discipline of the UNCED requirements, a discipline which is not usually placed on conferences, most of which are stand-alone. As a consequence, the plans for ICWE were conditioned to reflect the views of governments taking part in the preparations for UNCED. They also had to take inputs from a series of smaller meetings on various aspects of water resources, which were held in advance of the Dublin Conference.

1.9.2 The nature of ICWE

Conferences can have a wide variety of aims and a number of different formats, some with outputs of various kinds, others with none. In this case it was clear that ICWE had to be a conference where discussion of the wide range of water issues would result in the formulation of recommendations for action which could be discussed at Rio and implemented in the follow up to UNCED. Hence, ICWE was to follow a pattern different from the standard international scientific conference and even from many of the intergovernmental conferences held under the UN umbrella. It was to be convened by a group of 24 UN bodies and agencies, each with their own aspirations, but organized within the usual UN and agency rules of conduct. The agenda had to satisfy the needs expressed by the government delegates to the PrepComms, recognizing that the agenda had also to accommodate the water-orientated non-governmental and intergovernmental organizations. ICWE was meant to be for hydrologists, water engineers, economists, planners, ecologists, lawyers and other professionals active in the field of water and this view was endorsed at PrepComm 3. The delegates at that meeting wanted it to be a confer-

* UNDP Global Consultation on Safe Water and Sanitation for the 1990s, 10–14 September 1990. Safe Water 2000 *The New Delhi Statement.*

ence of government-appointed experts, rather than of government delegations, so that the position of governments would not be pre-empted and that governments represented at PrepComm 4 and UNCED would have the final say. This procedure avoided the danger that one government conference (ICWE) could have had its findings overturned at another government conference (PrepComm 4 or UNCED). This also meant that the recommendations from ICWE did not have to be directly incorporated into the UNCED Agenda 21. In the event, in PrepComm 4 and at Rio de Janeiro, some delegations argued against including the findings of ICWE and using them in Agenda 21 because ICWE was not a governmental conference. Such manoeuvres highlight the complexities of the organizational process and those of international politics.

1.9.3 Defining objectives and setting the agenda

The issues surrounding water resources management within environmental and developmental considerations are extremely complex. A strategy for effective discussion had to be devised which would focus the attention of participants on the most important issues (given that not everything could be addressed in a period of a few days at the Conference). Objectives had to be defined, issues had to be given priorities and there had to be a logic in the way they were to be discussed. This was an immensely difficult task and agreement on objectives and a basic agenda was very time-consuming.

A number of UN agencies (see 1.9.4 below) had responsibility for defining objectives and suggesting an agenda for discussion. The general objectives were agreed upon fairly readily; however, the format and agenda for discussion of issues at ICWE were strongly debated and agreement was difficult. It was only after the third UNCED PrepComm in September 1991 (only four months before ICWE) that the agenda for ICWE was finally agreed. This meant that countries and organizations to be involved could be given final instructions for preparation only at the last minute.

The main objectives of the Conference were:

(a) To assess the current status of the world's freshwater resources in relation to present and future water demands and to identify priority issues for the 1990s;
(b) To develop coordinated inter-sectoral approaches towards managing these resources by strengthening the linkages between the various water programmes;
(c) To formulate environmentally sustainable strategies and action programmes for the 1990s and beyond to be presented to the UNCED Earth Summit;
(d) To bring the above issues, strategies and actions to the attention of governments as a basis for national programmes and to increase awareness of the environmental consequences and developmental opportunities in improving the management of water resources.

The format for discussion was to be undertaken within six Working Groups which addressed:

(a) Integrated Water Resources Development and Management;
(b) Water Resources Assessment and Impacts of Climate Change on Water Resources;
(c) Protection of Water Resources, Water Quality and Aquatic Ecosystems;
(d) Water and Sustainable Urban Development and Drinking Water Supply and Sanitation in the Urban Context;
(e) Water for Sustainable Food Production and Rural Development and Drinking Water Supply and Sanitation in the Rural Context;
(f) Mechanisms for Implementation and Coordination at Global, National, Regional and Local Levels.

These working groups embraced the same topics as defined for structuring the freshwater section of the Agenda 21 document with the modification that the 'Impacts of climate change on water resources' and 'Drinking water supply and sanitation' were not treated as separate topics.

1.9.4 Responsibility for convening and organizing ICWE

ICWE was organized by the UN Administrative Coordination Committee – Intersecretariat Group for Water Resources (ACC-ISGWR). As the chairmanship of ISGWR, which rotates amongst the members, rested with the World Meteorological Organization (WMO) in 1991 and 1992, it fell to WMO to convene the Conference on behalf of the more than 20 UN agencies and bodies within the group. Thus, the Secretariat for the Conference was housed within WMO and was administratively responsible to WMO, while taking instruction from and acting for the ICWE Steering Committee, which was composed of representatives of ISGWR.

The ICWE Secretariat in WMO Geneva, reporting to the ICWE Steering Committee, was charged with coordination of all Conference activities. A Local Organizing Committee in Dublin, within the Department of the Environment of the Government of Ireland, was charged with organizing all the on-the-spot logistics and was in regular contact with the Secretariat in WMO.

While UN agencies acted behind the scenes putting together the conference organization, it was governments acting through the UNCED PrepComms which set the agendas and it was the government designated experts who took the major decisions in Dublin concerning the running of the Conference.

1.9.5 Funding

To cover the costs of organizing the ICWE it was estimated the sum of US$1.6m would be necessary. This sum was over and above the expenditures of the UN agencies and government departments in support of the salaries and expenses of individuals involved with the Conference.

Expenditures were incurred in three main areas in approximately equal measure. The salaries and expenses of members of the ICWE Secretariat had to be met. Some of these were met by individual governments loaning personnel to work on a 'no charge' basis. There were considerable costs incurred for the production, translation (into the six official UN languages) and distribution of documents. Lastly, funds were needed to finance the attendance of participants from less-developed countries.

To meet these costs funds were donated by the governments of a number of countries, by a number of the UN agencies most deeply involved with the issues addressed by the Conference and by several NGOs.

Two major problems were encountered. As ICWE was convened and organized at very short notice, there was not enough time to incorporate funds properly into the normal budgetary processes of the different UN and national bodies involved. This resulted in very great uncertainty about commitment of funds and it resulted, too, in several cases in the transfer of funds for use in the Conference at the very last moment. This caused a critical cash flow problem which made the functioning of the Secretariat particularly difficult. A second problem was that many agencies and governments were willing to provide funds for travel of participants from developing countries but not for other purposes. Thus, the support of the Secretariat became doubly difficult.

In the final analysis sufficient funds were raised to cover all major costs. Had this not been the case, the financial shortfall would have been a severe embarrassment to the ICWE Steering Committee.

1.9.6 Languages and documentation

At UN sponsored conferences it is standard procedure to translate documents into up to six languages and to provide simultaneous interpretation during the Conference itself in these languages. This can be a very costly and time-consuming process, especially if all six official UN languages are used*. With a very limited budget (with the final firm income not confirmed until after the Conference was over) and with a very tight time schedule for document production, it was clear from the start that it would be quite impractical to use all languages for all documents and for interpretation in all sessions of the Conference. In the year leading up to ICWE there was an ongoing discussion on which combination of languages to use. This was diplomatically very sensitive.

* The official UN languages are: Arabic, Chinese, English, French, Spanish and Russian.

It was finally agreed that English, French and Spanish would be the languages of interpretation, that key documents would be produced in six languages and that other documents would be translated from the original language of writing whenever possible.

1.9.7 The participants

It has already been stated that ICWE was to be primarily a conference of experts in water. It was realized, however, that to be effective the recommendations would have to be incorporated into the essentially political preparatory process of UNCED. It would therefore be practical to encourage a certain number of decision makers to participate in both ICWE and UNCED. There was always a certain vagueness about the exact nature of ICWE, a vagueness which led to a variety of interpretations of who should attend. This was to have ramifications for the appropriate conduct of procedures at the Conference.

There were concerns from the start as to how many participants would attend, from which countries and organizations they would come and whether there would be an appropriate geographical and organizational balance between participants. The venue chosen in Dublin could acceptably seat 500 people in conference style; more than 700 could not have been accommodated. Equally there would have been embarrassment if only 300 had attended. A timetable where three of the six Working Groups would meet simultaneously had been chosen for convenience and cost. Too many participants in any working group would render it ineffective. Too few participants would give less impact to its results and would adversely affect the viability of the Conference as far as hotel revenues were concerned.

It was not possible to invite individuals to attend. The strict protocol dictated that invitations had to be sent to Ministers of Foreign Relations within countries who would then inform whom they considered the most appropriate agencies or individuals from the country. Countries could not be told how many participants they could send. A maximum figure of six was suggested but it was proposed that ideally three or four should attend. There was a very real fear that developed countries would wish to send many participants and that there would consequently be an overwhelming preponderance of attendees from these nations.

The attendance of participants representing non-governmental and intergovernmental organizations also posed a very difficult problem. If a truly representative cross-section of viewpoints was to be achieved then many NGOs and IGOs should attend. However, there are literally hundreds of NGOs with interests in water and who was to judge which should or should not be invited to attend? In the end only international NGOs were encouraged to attend and only those officially accredited within the UNCED process were invited.

Special consideration had to be given to the attendance of VIPs. The Prime Minister of Ireland had kindly agreed to open the Conference and some of his ministers would participate in some of the formal ceremonies and press conferences. The executive heads of UN agencies would wish to attend and several government ministers from countries would want to participate. The strict rules of protocol and the uncertainty surrounding this matter until the last days before the Conference as to who would attend, posed tremendous challenges to the organizers.

Sponsorship of participants from developing countries was a question which was to be particularly frustrating and time-consuming for the Secretariat. It was straightforward to agree in principle that one participant from any developing country wishing to attend should be funded for travel and expenses. There were several factors which mitigated against easy decision making. As a result of the fragmentation of institutional jurisdiction over water and environmental matters, mentioned previously, it was often not clear within countries as to who was the most appropriate individual to attend. There were many instances in which more than one ministry claimed precedence. In such cases the ICWE Secretariat had no option but to refer the decision back to the country's ministry of foreign relations, a very time-consuming procedure. A second factor was that funding in support of participants came from several agencies and countries. Some of these transferred both their funds and the decision on whom to fund to the ICWE Secretariat. Several did not, preferring to choose which countries and in some cases which individuals they were willing to support. The Secretariat in these cases had to act as a clearing house, matching donors and recipients and attempting to maintain the guideline of support of one participant from each country.

In the event, there were 505 participants from 114 countries, 28 UN agencies and 58 NGOs and IGOs. Some 100 individuals from the media covered the Conference. The balance achieved was as near ideal as could reasonably be expected.

1.9.8 The ICWE Secretariat

A small Secretariat was established barely one year before ICWE was opened to coordinate all conference activities. It reported to the Chairman of the Conference Steering Committee and was housed within and received the full administrative support of WMO. It was financed from the Trust Fund established for the Conference through voluntary donations from UN agencies and from countries. Some personnel from the WMO Department of Hydrology and Water Resources were seconded to the Secretariat and several countries also seconded personnel to work within the Secretariat for a few months at a time. A Conference Coordinator was appointed to manage the Secretariat.

Within the Secretariat a special coordinator for Public Information and Publicity was appointed who organized, in conjunction with all the principal UN agencies involved, press conferences, promotion of ICWE through journals and all media activities. This very important activity took much time and effort.

The Secretariat had to liaise on a daily basis with members of the Steering Committee, with the UNCED Secretariat, with the Local Organizing Committee in Dublin and with all potential participants. The importance of a smooth working relationship between the Secretariat and WMO staff cannot be overemphasized.

1.9.9 Running the Conference in Dublin

ICWE lasted five and a half days in Dublin. After a separate Opening Ceremony and a day in which seven keynote papers were presented and a panel discussion was held to set the scene, all debate and decision making on recommendations for action had to be compressed into four days of work. The participants broke into three simultaneous Working Groups for two days for intensive discussion on particular issues and then reconvened in plenary to agree upon the final text of a *Dublin Statement* and a *Report of the Conference*. This was an incredibly tight schedule necessitating much forethought in the preparation of background documentation and strict adherence to agendas by the Chairs of the Working Groups.

So that the Conference would run smoothly and efficiently in a way acceptable to the participants, Conference officers were elected on the usual UN regional basis consisting of a Conference Chairman and Rapporteur, and Chairmen, Vice Chairmen and Rapporteurs for each of the six Working Groups, as shown in Table 6b. A smaller Conference Bureau (Table 6a), selected from those officers, met several times each day outside the normal sessions to oversee progress. Thus, while preparations leading to the Conference were guided by the ISGWR, it was the participants themselves who ran the process in Dublin.

There were many technical and logistic constraints. Of particular concern were documents produced for participants during the Conference itself. Between Working Group sessions summary documents would be written, sent to WMO in Geneva for overnight translation from English into French and Spanish and reproduced for distribution to participants before the start of the next session. The 'mechanical' problems of setting up copying machines and computers and the registering of hundreds of participants with their individual problems were relatively simple by comparison.

The greatest challenge was to allow participants to have freedom of expression and the ability to guide their own affairs, while at the same time giving the guidance necessary to achieve worthwhile results in a very short space of time.

Table 6(a). *ICWE Bureau structure*

Conference bureau
Conference Chairman – Mr James Dooge (Ireland) Conference Rapporteur – Ms Claudia Candanedo (Panama)
Conference Vice Chairmen – Mr Abbas Hidaytalla Abdullah (Sudan) Mr James Bruce (Canada) Mr Odon Starosolszky (Hungary) Mr Jose Luis Calderón (Mexico) Mr Chandra Sharma (Nepal) Mr Mahmoud Abu-Zeid (Egypt)
Mr J C Rodda (Chairman, ICWE Steering Committee) Mr G J Young (Coordinator, ICWE) Ms M Moylan (Department of the Environment, Ireland)

1.9.10 The outcomes

ICWE resulted in a short *Dublin Statement*, reproduced in Annex 1, and a *Report of the Conference*. Both these documents have been made available in the six official UN languages. The recommendations for action, which are detailed in the body of this book, were taken to the final PrepComm for UNCED, held in New York in March and April 1992. Many of the recommendations were incorporated into the UNCED Agenda 21 document, the blueprint for action into the twenty-first century.

Although the summary statements from ICWE have been presented in the *Report of the Conference*, much very valuable background material is contained in the background documents prepared to assist participants in Dublin, and in the various UN Strategy Documents and reports of meetings leading up to ICWE, many of which are not readily available. This book attempts to summarize material from all these documents, in order to present a more detailed and complete picture than that given, of necessity succinctly, in the formal output from ICWE.

ICWE was designed as the major lead in freshwater issues into the UNCED process. It was also a conference in its own right. Most of the recommendations from ICWE were incorporated into the Agenda 21 document. Many of the principles and recommendations for action put forward at ICWE are in the process of being adopted in countries in all regions of the world.

Table 6(b). *ICWE Working Group structure*

Working group	Chair	Vice Chair	Rapporteur
A. Integrated Water Resources Development and Management	Mr Abbas Hidaytalla Abdullah (Sudan)	Mr Mohamed Abdulrazzak (Saudi Arabia)	Mr Ainun Nishat (Bangladesh)
B. Mechanisms for Implementation and Coordination at the International, National and Local Levels	Mr James Bruce (Canada)	Mr Mohammed Jellali (Morocco)	Mr Abdul Karim Al-Fusail (Yemen)
C. Water Resources Assessment and Impacts of Climate Change on Water Resources	Mr Odon Starosolszky (Hungary)	Mr Karl Hofius (Germany)	Mr Moremi Sekwale (Botswana)
D. Protection of Water Resources, Water Quality and Aquatic Ecosystems	Mr Jose Luis Calderón (Mexico)	Mr Ivan Zavadsky (Czechoslovakia)	Ms LaJuana Wilcher (United States of America)
E. Water and Sustainable Urban Development and Drinking Water Supply and Sanitation in the Urban Context	Mr Chandra Sharma (Nepal)	Mr Gabriel Arduino (Uruguay)	Ms Ebele Okeke (Nigeria)
F. Water for Sustainable Food Production and Rural Development and for Drinking Water Supply and Sanitation in the Rural Context	Mr Mahmoud Abu-Zeid (Egypt)	Mr David Constable (Australia)	Mr Mario Fuschini-Mejía (Argentina)

1.9.11 Additional comments

It should be clear from this brief summary of activities that the order and logic behind conference organization apparent to the outsider belie the real situation. The organization of ICWE was done under extreme pressure, with an uncertain budget and with rules and agendas which were in a constant state of flux. Preparations by no means followed a logical order and it was only the common realization that tangible results were of the utmost importance that the Conference succeeded in achieving its primary goals in a largely acceptable form.

The necessity for integrated water resources management The malodorous outflow from a fish-processing plant pollutes slum dwellings in Latin America. Credit: WHO/UNICEF/D. Briggs.

2

Integrated water resources development and management

2.1 Background

The extent to which water resources development contributes to economic productivity and social well-being is not usually appreciated, although all social and economic activities rely heavily on the adequacy of the supply and quality of freshwater. As populations and economic activities grow, many countries are rapidly reaching conditions of water scarcity or facing limits to economic development. Water demands are increasing rapidly, with an estimated 70–80% required for irrigation, less than 20% for industry and a mere 6% for domestic consumption. The holistic management of freshwater as a finite and vulnerable resource, and the integration of sectoral water plans and programmes within the framework of national economic and social policy, are of paramount importance for action in the 1990s and beyond. The fragmentation of responsibilities for water resources development among sectoral agencies is proving, however, to be an even greater impediment to promoting integrated water management than had been anticipated. Effective implementation and coordination mechanisms are required (UN-DTCD/IBRD/UNDP, 1991; UN-DTCD, 1991b, c, d).

Institutional capacity for implementing integrated water management should be reviewed and developed. Existing administrative structures will often be quite capable of achieving local water resources management, but the need may arise for new institutions based upon the perspective of river catchment areas, district development councils and local community committees. Although water is managed at various levels in the socio-political system, demand-driven management requires the development of water-related institutions at appropriate levels, taking into account the need for integration with land-use management.

In creating the enabling environment for lowest-appropriate-level management, the role of Government includes mobilization of financial and human resources, legislation, standard-setting and other regulatory functions, monitoring and assessment of the use of water and land resources, and the creation of opportunities for public participation. International agencies and donors have an important role to play in providing support to developing countries in creating the required enabling environment for integrated water resources management. This should include donor support to local levels in developing countries, including community-based institutions, non-governmental organizations and women's groups.

The International Conference on Water and the Environment drew upon the large and diversified experience gained among all countries since the United Nations Water Conference, Mar del Plata, 1977. Some serious errors and unwise biases were identified, while at the same time there was recognition of numerous fruitful advances in understanding and action. There emerged four main principles that need to be applied in taking action to achieve integrated water resources development and management. The principles below are quoted directly from the Dublin Statement, the full text of which appears in Annex 1.

Principle No. 1 – Fresh water is a finite and vulnerable resource, essential to sustain life, development and the environment

Since water sustains life, effective management of water resources demands a holistic approach, linking social and economic development with protection of natural ecosystems. Effective management links land and water uses across the whole of a catchment area or groundwater aquifer.

Principle No. 2 – Water development and management should be based on a participatory approach, involving users, planners and policy-makers at all levels

The participatory approach involves raising awareness of the importance of water among policy-makers and the general public. It means that decisions are taken at the lowest appropriate level, with full public consultation and involvement of users in the planning and implementation of water projects.

Principle No. 3 – Women play a central part in the provision, management and safeguarding of water

This pivotal role of women as providers and users of water and guardians of the living environment has seldom been reflected in institutional arrangements for the development and management of water resources. Acceptance and implementation of this principle requires positive policies to address women's specific needs and to equip and empower women to participate at all levels in water resources programmes, including decision making and implementation, in ways defined by them.

Principle No. 4 – Water has an economic value in all its competing uses and should be recognized as an economic good

Within this principle, it is vital to recognize first the basic right of all human beings to have access to clean water and sanitation at an affordable price. Past failure to recognize the economic value of water has led to wasteful and environmentally damaging uses of the resource. Managing water as an economic good is an important way of achieving efficient and equitable use, and of encouraging conservation and protection of water resources.

2.2 Integrated water resources planning

Basis for action:

In all countries, the planning of water resources development is as important an activity as their efficient management. Effective planning is needed to solve the many problems inherent in the control and utilization of water: conflicting demands; too little or too much water; maximizing economic and social benefits; equity considerations; environmental and economic sustainability. The interdisciplinary nature of water problems requires new attitudes towards integrating the technical, economic, environmental, social and legal aspects into a coherent framework and the development and dissemination of planning methodologies, as demonstrated in OAS (1984).

Water resources development and management should be planned in an integrated manner, taking into account long-term planning needs as well as those with narrower horizons, that is to say, they should incorporate environmental, economic and social considerations based on the principle of sustainability; they should include the requirements of all users as well as those relating to the prevention and mitigation of water-related hazards; and they should constitute an integral part of the socio-economic development planning process (see Bamberger and Cheema, 1990). A prerequisite for the sustainable management of water as a scarce vulnerable resource is the obligation to acknowledge in all planning and development its full costs. Planning considerations should reflect on the one hand all types of benefit, both direct and indirect, and on the other all investment, environmental protection and operational costs, as well as the opportunity costs reflecting the most valuable alternative use of water. Actual charging need not necessarily burden all beneficiaries with the consequences of those considerations. Charging mechanisms should, however, reflect as far as possible both the true cost of water when used as an economic good and the ability of the communities to pay.

Strategy and programme targets:

The most effective way to promote the new approach is to apply it to all policy, programmes and project formulation exercises in order to assist governments in selecting appropriate strategies to meet considerations of sustainability, and human development. This is to be supported, in developing countries, by training and technical assistance.

Table 7. *Integrated water resources planning*

Activities and related means of implementation	Level[1]	Considered by	
		ICWE[2]	UNCED[3]
1. Diagnostic assessments in the water sector through rapid but comprehensive analyses of the existing status of water resources development, national goals and strategies, problems and priority areas for action.	N	X	X
2. National capacity building through: (a) training of water managers and professionals at all levels; (b) transfer of technology; (c) human resources development including the improvement of career structures; (d) institutional strengthening; (e) rationalization of public and private sector intervention; (f) development and strengthening of cooperation, including mechanisms, at all levels concerned, namely: – delegation of water resources management to the lowest appropriate level, including decentralization of government services to local authorities, private enterprises and communities and supporting water-user groups to optimize local water resources management; – at the national level, integrated water resources planning and management in the framework of the national planning process and establishment of independent regulation and monitoring of freshwater, based on national legislation and economic measures; – at the regional level, consideration of the harmonization of national strategies and action programmes; – at the global level, improved delineation of responsibilities, division of labour and coordination of international organizations and programmes, including facilitating discussions and sharing of experiences in areas related to water resources management.	INPL	X	X
3. Integrated information management through: (a) surveys of existing data; (b) assessment of needs and review of technology; (c) inventories of water resources, in combination with land-use planning, forest resource utilization, protection of mountain slopes and riverbanks and other relevant development and conservation activities; (d) development of interactive databases, forecasting models, economic planning models and methods for water management and planning, including environmental impact assessment methods; (e) formulation of data gathering programmes.	NP	X	X

4. Formulation of costed and targeted national action plans and investment programmes taking into account the optimization of water resources allocation under physical and socio-economic constraints and the need for the integration of water quantity and quality management. Such plans would include flood and drought management, risk analysis and environmental and social impact assessment. Plans should be monitored and evaluated and updated through the development of interactive and flexible machinery. Plans should be implemented at different scales including: (a) community development programmes and activities; (b) district and province programmes and projects; (c) multipurpose projects (including the special problems of man-made lakes); (d) river basin plans; (e) international watercourses.	NPL	X	X

[1] Level of implementation: I=International; N=National; P=Provincial or sub-national; L=Local.
[2] Considered in ICWE *Report of the Conference* section 2.
[3] Considered in UNCED Agenda 21 paragraph:18.6–22.

Table 8. *Demand management*

Activities and related means of implementation	Level[1]	Considered by	
		ICWE[2]	UNCED[3]
1. Water auditing to promote improvement in efficiency of supply and wastage minimization through: (a) the effective metering and measurement of volumes supplied; (b) leak detection and repair; (c) identification of illegal connections.	NPL	X	X
2. Development of water use policy to include water tariffs and other economic instruments to effect demand management in domestic supply, in agriculture and in industry.	N	X	X
3. Implementation of allocation decisions through demand management, pricing mechanisms and regulatory measures, taking into account: (a)legal and institutional aspects; (b) allocation of public expenditures; (c) accounting and auditing systems; (d) monitoring and evaluation.	NPL	X	X

[1] Level of implementation: I=International; N=National; P=Provincial or sub-national; L=Local.
[2] Considered in ICWE *Report of the Conference* section 2.
[3] Considered in UNCED Agenda 21 paragraph: 18.6–22.

All countries will be expected to have carried out by the year 2000 a diagnostic phase to develop a strategy and a planning phase aimed at costed and targeted national action plans.

2.3 Demand management

Basis for action:

Abundance or scarcity of water can mean prosperity or poverty, life or death. It can even be a cause of conflict. Most countries have serious problems concerning the quantity and quality of their freshwater resources. Constraints on the supply of fresh water are increasingly aggravated by droughts, depletion of aquifers, pollution and land degradation, while demand for water is rising rapidly for food production, industry and domestic consumption. A constraint of a different kind is the absence of detailed knowledge of the volumes of water 'used' by the different customers that would be obtained from the effective metering and measurement of supplies.

Pursuant to the recognition of water as a social and economic good, the various available options for charging water users (including domestic, urban, industrial and agricultural water-user groups) have to be further evaluated and field-tested, (see UN-DTCD, 1991a). Further development work is required for economic instruments that take into account opportunity costs and environmental externalities. Field studies on the willingness to pay should be conducted in rural and urban situations. Options for water conservation and reuse should be vigorously pursued. These ideas have been extensively reviewed by, for example, Borde and Pearce (1991), Pearce and Markandya (1989) and United Nations (1980).

Strategy and programme targets:

Rather than seeking a supply adequate for some set of water 'needs', water management is concerned with finding a balance between the benefits of water use and the costs of water supply. 'Needs' are no longer measured in consumption per capita per day, but in terms of the health and welfare of human populations. Costs are not linked to financial outlays for engineering and construction, but include all adverse effects on the economy, or activities which compete for the basic resources and on the environment.

Demand management should be introduced into all national action plans and implemented by the year 2000 (UN-DTCD, 1992a). The necessary training and transfer of technology should have taken place and at least half the developing countries should have carried out evaluation of the effectiveness of demand management.

Table 9. *Institutional arrangements*

Activities and related means of implementation	Level[1]	Considered by	
		ICWE[2]	UNCED[3]
1. Implementation of water and land resources management at the lowest appropriate level.	NPL	X	X
2. Creation of appropriate water authorities and coordination arrangements.	NP	X	X
3. Integration of water management at basin level.	INP	X	X
4. Inception of efficient and effective organizational alternatives for the provision of water-related public services and for operation and maintenance of projects.	NPL	X	X
5. Creation of international arrangements and organizations for planning, developing and protecting international waters.	IN	X	X

[1] Level of implementation: I=International; N=National; P=Provincial or sub-national; L=Local.
[2] Considered in ICWE *Report of the Conference* section 2.
[3] Considered in UNCED Agenda 21 paragraph: 18.6–22.

2.4 Institutional arrangements

Basis for action:

Sustainable water development is contingent on appropriate institutional arrangements. Such arrangements should ensure an unbiased and independent approach in policy making, planning, allocation, development, conservation, protection and in the monitoring and assessment of the water resources on which the other activities depend. They should also bring about optimum technical efficiency, and ensure effectiveness in the provision of water-related services.

Strategy and programme targets:

Centralized and sectoral approaches to water resources development and management have often proved inadequate in addressing local water management problems. Recognizing the need for a central mechanism capable of securing national economic and social interests, the role of government needs to change to enable the delegation of responsibility for water resources development and management to the most appropriate and efficient levels, including both the informal and formal private sectors.

Governments should have assessed their institutional arrangements and taken steps to establish more appropriate mechanisms as part of national action programmes by 1995.

2.5 Legal frameworks

Basis for action:

Policy decisions cannot be implemented successfully unless there is adequate water legislation. Based upon the agreed strategy to develop water resources, water legislation provides part of the enabling environment, ensuring as far as possible the most equitable, economic and sustainable use of available water resources. Such legislation is a complex endeavour since it has to take account of several simultaneous, and sometimes conflicting objectives: development objectives, including related public and private investments; environmental and conservation goals, requiring effective public control, but also demanding private sector cooperation and involvement; and social objectives, consisting mainly of water-related services and the social impact of development components.

At the international level effective treaties or joint or concurrent legislation are essential to deal with increasing instances of transboundary water pollution and conflicting demands on shared watercourse systems.

Table 10. *Legal frameworks*

Activities and related means of implementation	Level[1]	Considered by	
		ICWE[2]	UNCED[3]
1. Review and analysis of customary and existing water legislation.	NPL	X	X
2. Enactment of appropriate water resources legislation including regulations and by-laws.	NPL	X	X
3. Enactment of legally compulsory rules for the assessment of water projects and programmes.	NP	X	X
4. Enactment of legislation for the provision of water-related public services.	NPL	X	X
5. Bilateral, multi-lateral, regional and global international agreements on the use, environmentally-sustainable development, protection, and allocation of the resources of international water resources systems, with particular regard to transboundary water bodies.	IN	X	X

[1] Level of implementation: I=International; N=National; P=Provincial or sub-national; L=Local.
[2] Considered in ICWE *Report of the Conference* section 2.
[3] Considered in UNCED Agenda 21 paragraph: 18.6–22.

Strategy and programme targets:

Enactment of appropriate, enforceable and applicable legislation, both for water and for activities having an identified impact on water resources. Such legislation should at the same time encourage and enhance private sector participation and cooperation, and provide tools for expedient public intervention, when and as needed (all countries by the year 2000).

Global acceptance and effective application of rules of cooperation in good faith, environmentally sustainable management, equitable apportionment and prohibition of causing appreciable harm when developing and using the resources of international watercourse systems (acceptance of rules by 1995, application to large international watercourses by the year 2000).

2.6 Public participation

Basis for action:

No matter how efficiently the water resources planning and implementation process is carried out, its long-term impact and sustainability will depend on the effectiveness of public participation. This applies particularly to the full implementation of demand management, the establishment of a legal framework for water resources management and cost-recovery. In developing countries, the role of women in water resources management must be enhanced since they and their families are the prime users and beneficiaries of water development programmes and since they are often more concerned than men with the protection of the quality of surface- and groundwater (cf. Rodda, 1991).

The delegation of water resources management to the lowest appropriate level necessitates educating and training water management staff at all levels and ensuring that women participate equally in the education and training programmes. Particular emphasis has to be placed on the introduction of public participatory techniques, including enhancement of the role of women, youth, indigenous people and local communities. Skills related to various water management functions have to be developed by municipal government and water authorities, as well as in the private sector, local/national non-governmental organizations, cooperatives, corporations and other water-user groups. Education of the public regarding the importance of water and its proper management is also needed.

Strategy and programme targets:

A clear exposition to policy makers should be made of what is to be accomplished by involving the public in planning and management and how it can be achieved.

Table 11. *Public participation*

Activities and related means of implementation	Level[1]	Considered by	
		ICWE[2]	UNCED[3]
1.Promote public participation through: (a) development of extension courses; (b) public relations exercises including sharing of knowledge and technology; (c) dissemination of information to the public; (d) training in conflict resolution; (e) social impact assessments; (f) awareness-raising and educational programmes.	NPL	X	X
2. Promote community participation in planning, implementation, operation and maintenance, evaluation, monitoring.	PL	X	X
3. Enhance the role of women through: (a) participation in the decision-making process; (b) participation in projects and programmes; (c) development of training materials; (d) training of various target groups; (e) dissemination of results.	NPL	X	X

[1] Level of implementation: I=International; N=National; P=Provincial or sub-national; L=Local.
[2] Considered in ICWE *Report of the Conference* section 2.
[3] Considered in UNCED Agenda 21 paragraph: 18.6–22.

A major part of the strategy should be using the public information, education and training process to develop an iterative (i.e. top-down, bottom-up) open planning process; for example, training professionals in the sector in the use of the participatory techniques and applying the process to individual projects.

Since many of these objectives are difficult to quantify or specify as targets, self-evaluations by countries should be performed in the year 2000 to evaluate, at least qualitatively, the extent to which public participation has been enhanced and to gauge its impact on programme effectiveness.

2.7 Effective technologies

Basis for action:

To bring about the more effective integration of water resources development and management activities, a wide variety of technological options are available. These range from improved methods of data collection and handling, which enable the water resources planner to review different ways of developing a resource, to so-called 'non-conventional' methods of increasing the resource base, such as desalination and inter-basin transfer. The dissemination of knowledge of these techniques and options and the technology transfer needed to make them operational in developing countries is a priority area for action.

The development of interactive databases, forecasting methods and economic planning models appropriate to the task of managing water resources in an efficient and sustainable manner will require the application of techniques such as geographical information systems and expert systems to gather, assimilate, analyze and display multisectoral information and to optimize decision making. In addition, the development of new and alternative sources of water supply and low-cost water technologies will require innovative applied research. This will involve the transfer, adaptation and diffusion of new techniques and technology among developing countries, as well as the development of indigenous capacity, for the purpose of being able to deal with the added dimension of integrating engineering, economic, environmental and social aspects of water resources management and predicting their effects in terms of the human impact.

The setting afresh of priorities for private and public investment strategies should take into account (a) maximum utilization of existing projects, through maintenance, rehabilitation and optimal operation; (b) new or alternative clean technologies; and (c) environmentally and socially benign hydropower.

Table 12. *Effective technologies*

Activities and related means of implementation	Level[1]	Considered by	
		ICWE[2]	UNCED[3]
1. Incorporation of the concept of integrated water resources development and management into relevant university graduate and post-graduate courses.	NP	X	X
2. Giving priority support to technology transfer and national technical capacity building programmes and projects.	IN	X	X
3. Dissemination and diffusion of new and appropriate technologies to developing countries.	I	X	X
4. Promotion of international cooperation in scientific research on freshwater issues.	IN	X	X
5. Development of new and alternative sources of water supply such as sea-water desalination, artificial groundwater recharge, use of marginal-quality water, wastewater reuse and water recycling.	NP	X	X
6. Provision of venture capital for field testing of promising new technologies.	IN	X	X

[1] Level of implementation: I=International; N=National; P=Provincial or sub-national; L=Local.
[2] Considered in ICWE *Report of the Conference* section 2.
[3] Considered in UNCED Agenda 21 paragraph: 18.6–22.

Strategy and programme targets:

Developing countries need to strengthen their technological capabilities with the assistance of bilateral and multilateral organizations with regard to transfer of experience and know-how, technical cooperation and training.

Such technology transfer should be an integral part of the implementation national action plans, with the goal of reduced dependence on imported technologies and the establishment or strengthening of indigenous research and development facilities by the year 2000.

2.8 Targets and costs

(i) Targets

(a) All countries should have designed and initiated costed and targeted national action programmes and should have appropriate institutional structures and legal instruments in place by the year 2000;

(b) All countries should have established efficient water-use programmes to attain sustainable resource utilization patterns by the year 2000;

(c) Demand management should be introduced into all national action plans and implemented by the year 2000; the necessary training and transfer of technology should have taken place and at least half the developing countries should have carried out evaluations on the effectiveness of demand management;

(d) Self-evaluations by countries should be performed in the year 2000 to measure qualitatively the extent to which public participation has been enhanced and its impact on programme effectiveness;

(d) Sub-sectoral targets of all freshwater programme areas should have been achieved by the year 2025.

(ii) Cost estimates

During the period 1993 to 2000, an annual amount of about US$ 100 million of international financing is required to support national development in this programme area. The strengthening of international institutions in support of the planning and initiation phases at the country level requires the allocation of about US$ 10 million per year. Transboundary and global freshwater issues require a financial support in the order of US$ 5 million annually for the executing national, regional and global authorities and organizations. The total annual financing requirements in this programme area amount to about US$ 115 million from the international community on grant or concessional terms*.

* When estimating the financial requirements for integrated water resources management it has to be kept in mind that the bulk of investments and external donor support is covered under the programme areas on protection of water resources, urban water management and rural waste management.

Assessment of the resource is vital for understanding, planning and management. A meteorological measuring facility in West Africa. Credit: WMO.

3

Water resources assessment

3.1 Background

Water resources assessment (WRA), comprises the continuing determination of the location, extent, dependability and quality of water resources and of the human activities that affect those resources.

WRA is a prerequisite for sustainable development and management of the world's water resources. It provides the basis for the vast range of activities where water is involved. Without detailed WRA it is impossible properly to plan, design, construct, operate and maintain projects for the following purposes: irrigation and drainage; mitigation of flood losses; industrial and domestic water supply; urban drainage; energy production (including hydropower); health; agriculture; fisheries; drought mitigation and the preservation of aquatic ecosystems, estuarine and coastal waters.

The nature of the decisions based on WRA information may involve major capital investments with potentially massive environmental impacts. WRA provides a sound scientific basis for these investments, reducing risks and avoiding failure. It is also vital for government policies and programmes necessary to ensure sustainable development. Indeed, without the knowledge of the quantity and quality of surface- and groundwater resources derived from their comprehensive monitoring, it is impossible to determine whether sustainable development is being achieved and what to do if it is not.

There is growing concern that at a time when more precise and reliable information is needed about water resources, hydrological services and related bodies are less able than before to provide this information, especially information on groundwater and water quality. Major impediments are the lack of financial resources for water resources assessment, the fragmented nature of hydrological services and the insufficient numbers of qualified staff. At the same time, the

Table 13. *Institutional framework in support of water resources assessment*

Activities and related means of implementation	Level[1]	Considered by	
		ICWE[2]	UNCED[3]
1. Define the information needs of users for WRA, including the needs for flood and drought forecasting.	NP	X	X
2. Establish a national policy, a legislative framework, economic instruments and regulatory arrangements.	N	X	X
3. Establish the institutional arrangements needed to ensure the efficient collection, processing, storage, retrieval and dissemination to users of information about the quality and quantity of available water resources at the level of catchments and groundwater aquifers in an integrated manner.	N	X	X
4. Establish and maintain effective cooperation at the national level between the various agencies responsible for the collection, storage and analysis of hydrological data.	N	X	X
	N	X	X
5. Ensure that the assessment information is fully utilized in the development of water management policies.	N		X
6. Cooperate in the assessment of transboundary water resources.	IN	X	X
7. Encourage the application of methodologies developed and endorsed at the international level.	IN	X	
8. Develop and disseminate information on the means of estimating benefits and costs of WRA activities.	IN	X	

[1] Level of implementation: I=International; N=National; P=Provincial or sub-national; L=Local.
[2] Considered in ICWE *Report of the Conference* section 3.
[3] Considered in UNCED Agenda 21 paragraph: 18.23–34.

advancing technology for data capture and management is increasingly difficult for developing countries to access. Establishment of national databases is, however, vital to water resources assessment and to mitigation of the effects of floods, droughts, desertification and pollution.

3.2 Institutional framework in support of water resources assessment

Basis for action:

The assessment of the water resources of a country is a national responsibility and the activities concerned should be designed to meet the specific needs of that country. Many of its component activities may be undertaken at the local and provincial levels. Given the importance of assessment information to support sustainable development and for the maintenance of ecosystem integrity, all countries are urged to achieve a level of WRA activity appropriate to their needs as soon as is practicable.

Various institutional arrangements can support effective WRA programmes. In most countries responsibility for WRA is, unfortunately, divided between a number of ministries and national or provincial bodies. The growing need for integrated water resource management points to the desirability of close coordination between the different bodies involved in the collection, storage and analysis of the relevant data.

Should this responsibility be shared among neighbouring countries, such as in the case of transboundary water resources, international programmes and projects can provide valuable assistance.

Major international assessment initiatives have been undertaken through the Operational Hydrology Programme (WMO) and the International Hydrological Programme (UNESCO). These have been summarized for ICWE in WMO/UNESCO, (1991a) and Ayibotele, (1992).

Strategy and programme targets:

The national policy should be that all WRA activities are fully coordinated and adequately funded. The approach taken to achieve this may differ from country to country, but it will usually involve the establishment of regulations and a series of administrative decisions, particularly on the allocation of funds.

The success of these efforts can be measured by the general level of WRA activities and by a review of the degree of duplication of effort. Whether responsibility is centralized or distributed, WRA requires investment of financial resources if it is to provide the support to sustainable socio-economic development that is demanded of it. These resources, however, represent only a small fraction (say 0.2 to 1.0%) of the funds spent on the water sector as a whole.

Table 14. *Collection and storage of hydrological data*

Activities and related means of implementation	Level[1]	Considered by	
		ICWE[2]	UNCED[3]
1. Review existing data-collection networks and assess their adequacy, including those that provide real-time data for flood and drought forecasting.	IN		X
2. Install monitoring systems and improve networks designed to provide data on water quantity and quality for surface and groundwater, as well as relevant land-use data. Ensure the continuous operation of such systems in support of studies requiring long-term data, such as those relating to climate change.	N	X	X
3. Upgrade facilities and procedures used to store, process and analyze hydrological data and apply standards and other means to ensure data compatibility.	IN	X	X
4. Implement techniques for processing hydrological data and assimilating related information by: (a) establishing databases on the availability of all types of hydrological data at the national level; (b) implementing 'data rescue' operations, for example, by the establishment of national archives of water resources; (c) implementing appropriate well-tried techniques for the processing of hydrological data; (d) deriving area-related estimates from point hydrological data; (e) assimilating remotely sensed data and by using, where appropriate, geographical information systems.	IN	X	X
5. Make available to all countries water resources assessment technology that is appropriate to their needs, irrespective of their level of development, including methods for the assessment of the impact of climate change on freshwaters.	IN	X	X
6. Ensure the transfer of appropriate technology, particularly between hydrological services.	IN	X	

[1] Level of implementation: I=International; N=National; P=Provincial or sub-national; L=Local.
[2] Considered in ICWE *Report of the Conference* section 3.
[3] Considered in UNCED Agenda 21 paragraph: 18.23–34.

3.3 Collection and storage of hydrological data

Basis for action:

Reliable information on the condition of and trends in a country's water resources, including surface water, water in the unsaturated zone and groundwater, in respect of both quantity and quality, is required for a number of purposes, such as:

- assessing the resource and its potential for supplying the current and foreseeable demand
- protecting people and property against water-related hazards
- planning, designing and operating water projects
- monitoring the response of water bodies to anthropogenic influences, to climate variability and change, and to other environmental factors.

Integrated monitoring and information systems should be established and data collected and stored on all aspects of water resources which are required for a full comprehension of the nature of those resources and for their sustainable development. These include not only hydrological data, but also related geological, climatological, hydrobiological and topographic data and data on soil types, land use, desertification and deforestation, as well as data on water use and reuse, sewage discharges, point and non-point sources of pollution and the loads discharged to seas and oceans. This involves the installation of observation networks and other data-gathering mechanisms designed to monitor various climatic and topographic regimes, plus the development of data-storage and processing facilities. Where, at national, regional and international levels, water-related information is handled by a number of information systems, it is important that these systems be coordinated.

The capability of collecting and storing hydrological data varies markedly from country to country. Nowhere is it entirely satisfactory. Instrument networks may be too sparse, only a few of the hydrological variables may be measured and the records may be too short to be of value. In some cases the lack of data is extremely serious and poses a major problem for those planning long-term sustainable development.

Strategy and programme targets:

Data should be collected and stored on all aspects of water resources which are required for a full comprehension of the nature of those resources and for their sustainable development.

This involves the installation of networks of stations and other data-gathering mechanisms, the development of data-storage facilities and the systems for processing and analyzing the data in them.

Table 15. *Dissemination of water information*

Activities and related means of implementation	Level[1]	Considered by	
		ICWE[2]	UNCED[3]
1. Analyze and disseminate data and information on water resources in the forms required for planning and management of countries' socio-economic development and for use in environmental protection strategies and in the design and operation of specific water-related projects to assure the incorporation of water resources information in decision-making processes.	N	X	X
2. Disseminate assessments of the risk of flooding from rainfall, snowmelt, storm surges and land-slides by installing hydrological forecasting and warning systems within the context of the International Decade for Natural Disaster Reduction (IDNDR).	N	X	X
3. Disseminate assessments of the risk of drought by installing drought warning systems in support of schemes to mitigate the effects of drought within the context of the IDNDR.	N	X	X
4. Disseminate assessments of surface-water and groundwater resources and of the interactions between surface water and groundwater.	N	X	
5. Disseminate basin-wide, regional and global sets of water-related data and information for use, *inter alia*, in the management of resources within international river basins and in climate change studies.	IN	X	

[1] Level of implementation: I=International; N=National; P=Provincial or sub-national; L=Local.
[2] Considered in ICWE *Report of the Conference* section 3.
[3] Considered in UNCED Agenda 21 paragraph: 18.23–34.

3.4 Dissemination of water information

Basis for action:

The collection of hydrological data is not an end in itself. To have value, data must be used and their use must have an impact on decisions. This does not exclude the possibility that some data are collected as an insurance against future unforeseen needs.

Nevertheless, too often the potential user of hydrological data is not considered in the planning of data-collection systems and is not adequately informed of the availability of data that are collected.

Those who plan, design and operate water projects, and those who are concerned with the protection of life, property and the environment from natural or man-made disasters, should have access to the water-related information necessary for their work. They should be informed of the availability of such information and be able to obtain it in forms that are convenient for their use, including the free and urgent exchange of data required for mitigating natural disasters.

To be of real value in practical work, data must be compiled into sets and disseminated in a form appropriate for their use. There is a very real and growing need for large-scale regional and global sets of hydrological data for use in studies of global change, in particular within the context of climate change.

The approach should be to assess the data and information needs of potential users and to match these with the services provided by information centres and forecasting systems. This includes the strengthening of existing global data bases and the call for countries to supply data to such bases. In this, increasing use will be made of geographic information systems and similar computer-based technology.

Commercialization of water-related information should not prevent its full use, and dissemination of water-related information should be on a non-profit basis.

One particular application of hydrological data, and one that is being highlighted in the 1990s during the International Decade for Natural Disaster Reduction (IDNDR), is in the installation and operation of the hydrological forecasting systems, which are vital to safeguard lives and property in the face of major natural disasters.

Strategy and programme targets:

Those who plan, design and operate water projects, and those who are concerned with the protection of life, property and the environment from natural or man-made disasters, should have access to the water-related information necessary for their work. They should be informed of the availability of such data and be able to obtain them in forms that are convenient and timely for their use.

Table 16. *Research and development in the water sciences*

Activities and related means of implementation	Level[1]	Considered by	
		ICWE[2]	UNCED[3]
1. Establish and strengthen research and development programmes at national, regional and international levels so as to increase understanding of the fundamental processes involved in the water cycle, including the interactions between water, land and the atmosphere, and to support WRA and hydrological forecasting activities.	IN	X	X
2. Promote the development of new technology for WRA and hydrological forecasting.	IN	X	
3. Monitor research and development activities to ensure that they make full use of local expertise and other local resources.	IN	X	X
4. Transfer appropriate technology to users.	IN	X	

[1] Level of implementation: I=International; N=National; P=Provincial or sub-national; L=Local.
[2] Considered in ICWE *Report of the Conference* section 3.
[3] Considered in UNCED Agenda 21 paragraph: 18.23–34.

The approach is to assess the data needs of potential users and to match these with the services provided by information centres and forecasting systems.

3.5 Research and development in the water sciences

Basis for action:

WRA (including studies of floods, drought and desertification, and the use of hydrological forecasting) should be based on a sound understanding of the scientific principles involved, which in turn are dependent on technology for their implementation. Both science and technology have made remarkable progress in recent years, but there are still large areas where the water sciences have yet to make a significant breakthrough and where new technological developments are badly needed. These concerns have been addressed in National Research Council (1991), Ayibotele and Falkenmark (1992) and Plate (1992).

In addition to a lack of resources, research and development in the hydrological sciences suffer from insufficient coordination and a need to take more account of regional and national variations in the problems to be solved and in the expertise available. Research and development activities should necessarily be based on a strategic analysis of the very varied needs of countries. They should take account of, and strengthen, indigenous expertise.

Important research needs include (a) development of global hydrological models in support of analysis of climate change impact and of macroscale water resources assessment; (b) closing the gap between terrestrial hydrology and ecology at different scales, including the critical water-related processes behind loss of vegetation and land degradation and its restoration; and (c) study of the key processes in water-quality genesis, closing the gap between hydrological flows and biogeochemical processes. The research models should build upon hydrological balance studies and also include the consumptive use of water. This approach should also, when appropriate, be applied at the catchment level.

Water resources assessment necessitates the strengthening of existing systems for technology transfer, adaptation and diffusion, and the development of new technology for use under field conditions, as well as the development of endogenous capacity. Prior to inaugurating the above activities, it is necessary to prepare catalogues of the water resources information held by government services, the private sector, educational institutes, consultants, local water-use organizations and others.

Strategy and programme targets:

Research and development activities should be planned so as to meet the very varied needs of countries and should take account of indigenous expertise.

Table 17. *Human resources development*

Activities and related means of implementation	Level[1]	Considered by	
		ICWE[2]	UNCED[3]
1. Ensure sufficient numbers of appropriately qualified and capable staff are recruited and retained by water resources assessment agencies and provided with the training and retraining they will need to carry out their responsibilities successfully.	N	X	X
2. Identify education and training needs geared to the specific requirements of countries.	IN	X	X
3. Establish and strengthen education and training programmes on water-related topics, within an environmental and developmental context, for all categories of staff involved in WRA activities, using advanced educational technology where appropriate and involving both men and women.	INP	X	X
4. Develop sound recruitment, personnel and pay policies for staff of national and local water agencies.	NPL	X	X
5. Strengthen the managerial capabilities of water-user groups, including women, youth, indigenous people and local communities, to improve water-use efficiency at the local level.	NPL		X
6. Use water projects for on-the-job training of local staff.	NPL		
7. Promote the national capacity for the organization of workshops, seminars and conferences on subjects related to WRA and flood forecasting.	N		

[1] Level of implementation: I=International; N=National; P=Provincial or sub-national; L=Local.
[2] Considered in ICWE *Report of the Conference* section 3.
[3] Considered in UNCED Agenda 21 paragraph: 18.23–34.

Great benefit can be gained from a cooperative approach at both the national and international levels.

3.6 Human resources development

Basis for action:

Salaries and wages commonly account for half or more of the expenditure of an effective programme for WRA and hydrological forecasting. People are the most important resource available to the manager of such a programme, and because of this, personnel matters should receive great attention. These matters include the assessment of personnel requirements, the provision of attractive terms of employment and the establishment and use of effective schemes for the education and training of staff. Human resources development, an integral part of Capacity Building is treated in detail in IHE-Delft/UNDP (1991) and in Chapter 9 of this book.

Water resources assessment requires the establishment and maintenance of a body of well-trained and motivated staff sufficient in number to undertake the above activities. Education and training programmes designed to ensure an adequate supply of these trained personnel should be established or strengthened at the local, national, sub-regional or regional level. In addition, the provision of attractive terms of employment and career paths for professional and technical staff should be encouraged. Human resource needs should be monitored periodically, including all levels of employment. Plans have to be established to meet those needs through education and training opportunities and international programmes of courses and conferences.

Because well-trained people are particularly important to water resources assessment and hydrological forecasting, personnel matters should receive special attention in this area. The aim should be to attract and retain personnel to work on water resources assessment who are sufficient in number and adequate in their level of education to ensure the effective implementation of the activities that are planned. Education may be called for at both the national and the international level, with adequate terms of employment being a national responsibility.

The conduct of water resources assessment on the basis of operational national hydrometric networks requires an enabling environment at all levels.

Strategy and programme targets:

The aim should be to attract and retain personnel to work on WRA who are sufficient in number and adequate in their level of education to ensure the effective implementation of the activities that are planned.

Education may be called for at both the national and international level, while adequate terms of employment are a national responsibility.

3.7 Targets and costs

Targets

(a) All countries should have studied in detail the feasibility of installing WRA services by the year 2000;
(b) There should be WRA services with a high-density hydrometric network installed in 70 countries, and services with limited but adequate capacity in 60 additional countries, by the year 2000;
(c) There should be 110 countries with fully developed services, and 40 additional countries with services of a limited capacity, by the year 2025;
(d) The longer-term target is to have fully operational services, based upon high-density hydrometric networks, available in all countries.

Cost estimates

In order to attain the targets for the year 2000, total average annual funding (1993–2000) in the order of US$ 355 million is required, including contributions from external sources of US$ 145 million on grant or concessional terms*. The strengthening of international institutions for the development and exchange of information and technology requires about US$ 5 million per year (included above).

* Financial support is required primarily for the establishment and strengthening of national hydrometric networks, also covering transboundary watercourses.

4

Protection of water resources, water quality and aquatic ecosystems

Protection of water quality for human consumption is essential Purification plants like this one in Europe to counteract chemical pollutants in river or lake water are essential – but costly. Credit: WHO/Inter Nationes.

4

Protection of water resources, water quality and aquatic ecosystems

4.1 Background

As virtually every molecule of water on and under the Earth's surface is linked through the hydrological cycle and the total amount of water has a finite limit, the freshwater portion is often referred to as a unitary resource. Consequently, long-term sustainable development of global fresh water requires holistic management of the resource and recognition of the inter-connectedness of the elements that comprise fresh water and impact on its quality. Groundwater and surface water quality are inextricably linked. There is a growing recognition that issues of water quality cannot be considered separate from water quantity. The traditional emphasis on chemical indicators of water quality must be supplemented by more comprehensive indicators based on the total properties of a water body, including the physical, chemical, biological, radiological and ecological parameters of the water and of the material it carries. A global treatment of this subject is provided by Meybeck *et al.* (1989).

It must also be recognized that freshwater quality is impacted directly by natural and human activities outside the water sphere, such as land-use practices, erosion, and deforestation. Sources of pollution include both (a) point sources, for example effluent from an industrial concern, and (b) non-point sources, for example as a result of widespread spraying of crops with pesticides. Certain water quality problems are tied to acid deposition or natural contamination. Some problems require monitoring and protection at the local level, while others have significant transboundary components which can be addressed only at the national and international levels. All in all, the complex inter-connected nature of the freshwater system demands that freshwater management be holistic rather than piece-meal, systematically based rather than micro-managed, and based on a balanced consideration of the total needs of the people and the environment. These concepts are further developed by Falkenmark and Lundqvist (1992).

The objective of holistic management will be more likely to be achieved if values, direct and indirect, obvious and intrinsic, are calculated and the true cost of water is paid by the user. The payment of the true cost of water is likely to encourage water conservation, efficiency and reuse. It is being recognized more and more that efficient water use and the reuse of waste water are the most cost-effective and environmentally sound ways to address water supply needs. The ability-to-pay of the poor urban and rural people should be considered in pricing water supplies for essential uses.

A basin approach to water resources planning and management means considering all sources of pollution, point and non-point, including acid deposition and the leaching of contaminants from the soil into groundwater. It means addressing the connections between surface water and groundwater. It means considering the relationship between water quality and water quantity, and between upstream and downstream uses. And it means protecting and, as appropriate, restoring the chemical, physical and biological characteristics of water systems.

Several approaches, tools and mechanisms are recommended to implement holistic basin management. As a general approach, preventing pollution through reduced loadings or better management practices is preferable, both economically and environmentally, to cleaning up water resources after they are fouled. Scientific research, analysis, monitoring, surveillance, forecasting, prediction and assessment are important tools on which to rely when making management or development decisions. Environmental, social, health, economic, technical and legal considerations all must be taken into account and applied in an appropriate balance, when making management decisions and when implementing them.

There are few regions of the world that are still exempt from problems of loss of potential sources of freshwater supply, degraded water quality and pollution of surface and groundwater sources. Major problems affecting the water quality of rivers and lakes arise, in variable order of importance according to different situations, from for example: inadequately treated domestic sewage; inadequate controls on the discharges of industrial waste waters; ill-considered siting of industrial plants; deforestation; uncontrolled shifting cultivation; and poor agricultural practices. Some of these practices give rise to the leaching of nutrients and pesticides. Others disturb aquatic ecosystems and threaten living freshwater resources. Under certain circumstances, aquatic ecosystems are also affected by water resource development projects such as dams, river diversions, water installations and irrigation schemes. Erosion, sedimentation, deforestation and desertification have led to increased land degradation, and the creation of reservoirs has, in some cases, resulted in adverse effects on ecosystems. Many of these problems have arisen from a development model that is environmentally destructive and from a lack of public awareness and education about surface and groundwater resource protection. Ecological and human health effects are the measurable consequences, although the means to monitor them are inadequate or non-existent in many countries. There is a widespread lack of perception of the linkages between the development, management, use and treatment of water resources and aquatic ecosystems. A preventive approach, where appropriate, is crucial to the avoidance of costly subsequent measures to rehabilitate, treat and develop new water supplies.

The extent and the severity of contamination of unsaturated zones and aquifers have long been underestimated owing to the relative inaccessibility of aquifers and the lack of reliable information on aquifer systems. The protection of groundwater is therefore an essential element of water resource management.

Objectives

Three objectives will have to be pursued concurrently to integrate water-quality elements into water resource management:

(a) Maintenance of ecosystem integrity, according to a management principle of preserving aquatic ecosystems, including living resources, and of effectively protecting them from degradation on a drainage basin basis;
(b) Public health protection, a task requiring not only the provision of safe drinking-water but also the control of disease vectors in the aquatic environment;
(c) Human resources development, a key to capacity-building and a prerequisite for implementing water-quality management.

The effective protection of water resources and ecosystems from pollution requires considerable upgrading of most countries' present capacities. Water quality management programmes require a certain minimum infrastructure and staff to identify and implement technical solutions and to enforce regulatory action. One of the key problems today and for the future is the sustained operation and maintenance of these facilities. In order not to allow resources gained from previous investments to deteriorate further, immediate action is required in a number of areas.

Innovative approaches should be adopted for professional and managerial staff training in order to cope with changing needs and challenges. Flexibility and adaptability regarding emerging water pollution issues should be developed. Training activities should be undertaken periodically at all levels within the organizations responsible for water quality management and innovative teaching techniques adopted for specific aspects of water quality monitoring and control, including development of training skills, in-service training, problem-solving workshops and refresher training courses.

Suitable approaches include the strengthening and improvement of the human resource capabilities of local government in managing water protection, treatment

Table 18. *Water resources protection and conservation*

Activities and related means of implementation	Level[1]	Considered by	
		ICWE[2]	UNCED[3]
1. Preparation of national plans for water resources protection and conservation which would include the establishment of inter-sectoral programmes and coordination mechanisms to integrate land-use planning with water use and conservation requirements.	N	X	X
2. Development and application of water quality and water supply criteria for ecosystems and health protection.	INPL	X	
3. Preparation of basin action plans, especially for priority high-risk basins including their rivers, lakes, and aquifers. These plans should integrate land-use planning with water management and conservation, and there should be coordination of activities of provincial, national and, where necessary, international agencies. Contingency plans should be developed to control accidental spills and to respond to natural disasters.	INP	X	
4. Establishment of permanent, multi-sectoral planning and environmental impact assessment processes for water resources development and management covering hydrological, ecological, social, health, economic and meteorological aspects. These processes should be used at the national and international levels in government agencies and external support agencies and they should be simple and readily applicable.	IN	X	
5. Establishment and strengthening of appropriate legislation, enforcement and economic mechanisms for water resources protection and conservation at the national level with international cooperation to promote water conservation and recycling, pollution prevention and control, and environmentally-sound agricultural and industrial practices.	IN	X	X
6. Provision, where appropriate, of tax incentives and other economic benefits to industry for water conservation, recycling of water and minimization of waste discharges. Subsidizing environmentally-sound agricultural practices and compensating for restrictions needed to reduce water pollution from diffuse agricultural sources.	N		
7. Establishment and strengthening of technical and institutional capacities to identify and protect potential sources of water supply within all sectors of society.	N		X

[1] Level of implementation: I=International; N=National; P=Provincial or sub-national; L=Local.
[2] Considered in ICWE *Report of the Conference* section 4.
[3] Considered in UNCED Agenda 21 paragraph: 18.35–46.

and use, particularly in urban areas, and the establishment of national and regional technical and engineering courses. The subjects of suitable courses would be: water quality protection and control (at existing schools) together with courses on water resources protection and conservation for laboratory and field technicians, women and other water-user groups.

Governments should undertake cooperative research projects to develop solutions to technical problems that are appropriate for the conditions in each watershed or country. Governments should consider strengthening and developing national research centres linked through networks and supported by regional water research institutes. The North–South twinning of research centres and field studies by international water research institutions should be actively promoted. It is important that a minimum percentage of funds for water resource development projects is allocated to research and development, particularly in externally funded projects.

4.2 Water resources protection and conservation

Basis for action:

Growing demands for water have brought increased pressures on finite supplies. An ecosystem approach is necessary to provide water of adequate quantity and quality to all users and to protect ecosystem integrity over the long term.

There is a widespread lack of perception of the linkages between the development, management, use and treatment of water resources and aquatic ecosystems. A preventive approach is crucial to avoid costly subsequent measures to rehabilitate and treat existing water supplies and to develop new ones.

The relative scarcity of water resources in the face of continuously growing demands requires rational and sustainable management of this limited resource. The development of water resources implies their protection from overexploitation, pollution and degradation. Limited resources demand conservation with regard to parsimonious use as well as conservation in terms of quantity, quality and ecosystem integrity.

Strategy and programme targets:

Protection and conservation measures should be derived from the understanding of the aquatic environment as a coherent ecosystem which is integrated with other terrestrial ecosystems. Land-use activities, agricultural practices and industrial developments must be planned, taking water resource requirements into account, with regard to water supply demands and pollution impacts. A preventive approach is crucial to avoid costly subsequent containment and rehabilitation measures.

Table 19. *Monitoring and surveillance of water resources*

Activities and related means of implementation	Level[1]	Considered by	
		ICWE[2]	UNCED[3]
1. Establishment of networks for the monitoring and continuous surveillance of waters which receive wastes and of point and diffuse sources of pollution. The networks should take into account both surface and ground waters, water quality and water quantity and address all pollution types. Surveillance of pollution sources are undertaken to improve compliance with standards and regulations. Monitoring and assessment of complex aquatic systems often require multi-disciplinary studies involving several institutions and scientists in joint programmes.	NPL	X	X
2. The assessments should be harmonized within a basin framework (station networks, field and laboratory techniques, methodologies and procedures, data handling), leading to basin-wide data systems.	INP	X	
3. Water quality monitoring should be integrated with land-use practices, socio-economic developments and other environmental compartments. Land use should be rationalized to prevent land degradation and erosion and siltation of lakes and other water bodies.	NPL	X	X
4. The establishment or enhancement of national or regional water quality centres to ensure quality control and allow valid intercomparisons within and between basins and at international levels.	IN	X	
5. Promotion of new assessment and prediction techniques and methodologies such as low-cost field measurements, continuous and automatic monitoring, use of biota and sediment for micro-pollution monitoring, remote sensing, geographic information systems and Global Resource Information Database (GRID) methods.	IN	X	
6. Establishment and enhancement of effective flood and drought warning and preparedness systems within the framework of the International Decade for Natural Disaster Reduction.	IN	X	

[1] Level of implementation: I=International; N=National; P=Provincial or sub-national; L=Local.
[2] Considered in ICWE *Report of the Conference* section 4.
[3] Considered in UNCED Agenda 21 paragraph: 18.35–46.

By the year 2000 all countries should have identified those surface and groundwater resources which could be developed for use on a sustainable basis and other major water-dependent resources which can be developed. Simultaneously, they should have initiated programmes for the protection, conservation and rational use of these resources.

4.3 Monitoring and surveillance of water resources

Basis for action:

Monitoring, assessment, forecasting and prediction of the quality and quantity of rivers, lakes and groundwaters through water, biota and sediment is a goal for sound water resources management and protection. In addition to providing wildlife habitat, protecting and improving water quality and offering flood protection, aquatic systems are also linked to all other environmental components, and are therefore a powerful indicator of the overall environmental quality. Exact, complete and precise water quality data are needed from the local level to the international level for transboundary water bodies, and to the global scale where rivers provide a major input of pollutants to seas and oceans.

Water quality management requires an information base. Without one, rational policy formulation and decision making is not possible in an economically oriented framework for integrated water resources management. Compliance monitoring is needed to enforce water quality standards and effluent permits.

Strategy and programme targets:

The prime objective is to strengthen national and international capabilities for the establishment and operation of adequate water quality monitoring networks, including improved staff skills and institutional mechanisms.

Assessment of regionally or globally significant issues, including land-based pollutant fluxes to the oceans, long-range atmospheric transport of pollutants and land degradation, erosion, etc.

By the year 2000 all countries should have established appropriate assessment programmes of their water resources, and should participate in regional (e.g. basin level) and international water quality and quantity assessments such as the Global Water Quality Monitoring Programme GEMS/WATER.

Table 20. *Water pollution prevention and control*

Activities and related means of implementation	Level[1]	Considered by	
		ICWE[2]	UNCED[3]
1. Development and application of rapid assessment procedures for the identification and quantification of pollution sources; effluent monitoring; monitoring of acid precipitation; industrial and municipal sector reviews; and agrochemical use verification. Development of the use of risk assessment and risk management for decision making.	NPL	X	X
2. Establishment of standards for the discharge of effluents and for the receiving waters; identification and application of the best environmental practices at reasonable cost.	NP		X
3. Development of national and international legal instruments required to protect the quality of water resources, particularly for: (a) monitoring and control of pollution and its effects in national and transboundary waters; (b) control of long-range atmospheric transport of pollutants; (c) control of accidental and deliberate spills in national and transboundary water bodies.	IN		X
4. Enactment of water-specific or comprehensive environmental laws which cover preventive measures and pollution source control, including regulations and water quality standards.	N		
5. Introduction of the precautionary approach in water quality management, with a focus on pollution minimization and prevention through use of new technologies, product and process change, pollution reduction at source and effluent reuse, recycling and recovery, treatment and environmentally safe disposal.	NPL		X
6. Strengthening and enforcement of pollution prevention and control measures.	NP	X	
7. Utilization of economic instruments, including: (a) charges on water users; (b) application of the 'polluter pays' principle; (c) development of incentives; (d) mandatory environmental impact assessment of all major water resource development projects.	NPL	X	X
8. Development of programmes in priority areas of high risk for the restoration and enhancement of degraded aquatic ecosystems.	NPL	X	

Action	Level[1]	ICWE[2]	UNCED[3]
9. Promotion of national legislation and regional agreements for preventing and controlling transboundary water pollution.	IN	X	X
10. Protection of public health by the development of programmes for the identification and control of disease vectors and pathogens transmitted through fresh water.	IN	X	
11. Development and application of appropriate, clean, low-cost, low-waste technology for industrial production and sewage treatment, recycling of wastewater, and biotechnology for waste treatment taking into account indigenous technologies for water pollution prevention and control. Promotion of the construction of treatment facilities for domestic sewage and industrial effluents. Treatment of municipal waste water for safe reuse in agriculture and aquaculture.	IN	X	X

[1] Level of implementation: I=International; N=National; P=Provincial or sub-national; L=Local.
[2] Considered in ICWE *Report of the Conference* section 4.
[3] Considered in UNCED Agenda 21 paragraph: 18.35–46.

Table 21. *Protection of groundwater from pollution*

Activities and related means of implementation	Level[1]	Considered by	
		ICWE[2]	UNCED[3]
1. Identification and quantification of major pollution sources. Establishment of national inventories (including mapping programs) of groundwater resources and their character and determination of their responses to development activities. Such aquifer information will permit water managers to identify recharge and abstraction areas and interactions between surface waters and aquifers and establish controls on the types of activities which take place in these zones.	NPL	X	
2. Monitoring of surface- and groundwaters potentially affected by sites storing toxic and hazardous materials.	NPL		X
3. Ensuring that resource management and legislation provides for the sustainable management of groundwater, and promoting the development of national legislation for protecting groundwater.	N	X	X
4. Promotion of conservation and environmentally sound practices, including: (a) development of agricultural practices that do not degrade groundwaters; (b) design and management of landfills based upon sound hydrogeological information and impact assessment, using the best practicable and best available technology; (c) promotion of measures to improve the safety and integrity of wells and well-head areas to reduce intrusion of biological pathogens and hazardous chemicals into aquifers at well sites.	NPL	X	X
5. Encouragement and development of technologies for the promotion of waste minimization, pre-treatment and recycling.	NP	X	
6. Protection of groundwater abstraction and recharge areas through the establishment of protective zones where polluting activities are prohibited.	NPL		
7. Application of the necessary measures to mitigate saline intrusion into aquifers of small islands and coastal plains as a consequence of sea level rise or overexploitation of coastal aquifers.	NP	X	X

[1] Level of implementation: I=International; N=National; P=Provincial or sub-national; L=Local.
[2] Considered in ICWE *Report of the Conference* section 4.
[3] Considered in UNCED Agenda 21 paragraph: 18.35–46.

4.4 Water pollution prevention and control

Basis for action:

Despite the many efforts to contain water pollution and maintain water quality, pollution is having a growing impact on aquatic ecosystems: the past decade has witnessed a general deterioration in the quality of most of the vital water resources on all continents. Water availability and its suitability for different uses in industry, agriculture and for domestic supply is co-determined by its quantity and quality. This makes water quality protection a mandatory component of integrated water resources management.

Strategy and programme targets:

Three objectives have to be pursued concurrently in water quality management:

(a) Maintenance of ecosystem integrity and the protection of aquatic resources from pollution and its impacts due to socio-economic development;
(b) Public health protection through the control of disease vectors and pathogens in water resources;
(c) Sustainable water use by providing adequate amounts of water of a suitable quality on a long-term basis.

In order to slow down the rapid deterioration of water quality and enhance the availability of safe water, by the year 2000 all countries should have in place water pollution control programmes based upon enforceable standards for major point-source discharges, as well as for major non-point sources of pollution. These should include an inventory of potential sources of water supply leading to the preparation and implementation of programmes for their protection, conservation and sustainable utilization.

4.5 Protection of groundwater from pollution

Basis for action:

The extent and severity of contamination of the unsaturated and the saturated zones has long been underestimated due to the relative inaccessibility of aquifers and the lack of reliable information on aquifer systems. Pollutants are leaching into groundwater bodies from on-site disposal of excreta, landfill sites and mines. Fertilizers and pesticides are infiltrating into the groundwaters underlying agricultural areas. Many coastal and island areas are suffering from salt water intrusion as a result of overpumping. Preventative measures have become a matter of urgency since groundwater bodies, once contaminated, take a long time to recover from pollution and will thus be unavailable for vital supply needs for many years.

Table 22. *Protection of aquatic ecosystems and living freshwater resources*

Activities and related means of implementation	Level[1]	Considered by	
		ICWE[2]	UNCED[3]
1. Planning and implementing environmentally sound management of aquatic and terrestrial ecosystems including catchment and riparian forests, wetlands, riverine floodplains and associated freshwater and estuarine habitats as integral components of comprehensive water resources development.	INP	X	
2. Mandating environmental impact assessments for all major projects in basins considering social, health, economic and ecological concerns.	NPL	X	
3. Planning and implementation of pollution control programmes with particular attention to the protection of aquatic habitats and inland fisheries.	NP		
4. Rehabilitation of polluted and degraded lands, surface water and groundwater bodies to restore aquatic habitats, maintain biodiversity and improve the quality of the water for various human uses.	PL		X
5. Initiation of ecotoxicological studies of the long-term effects of particularly harmful chemicals and noxious aquatic species on aquatic biota and related ecosystems.	IN		
6. Maintaining, restoring or enhancing the ecological productivity and diversity of the wetland ecosystems which are important for their social, economic and environmental values.	PL	X	X
7. Urging parties to the Convention on Wetlands of International Importance Especially as Waterfowl Habitat (RAMSAR Convention) to apply its provisions in support of its environmentally protective recommendations and to encourage non-contracting parties to join in the convention.	IN	X	

[1] Level of implementation: I=International; N=National; P=Provincial or sub-national; L=Local.
[2] Considered in ICWE *Report of the Conference* section 4.
[3] Considered in UNCED Agenda 21 paragraph: 18.35–46.

Strategy and programme targets:

A strategy for the protection of groundwater must be aimed at protecting aquifers from becoming contaminated and preventive efforts should be directed first at land-use activities that pose a high risk of causing pollution from both point and non-point sources. Care must be exercised to avoid groundwater development that leads to the degradation of groundwater quality or the depletion of groundwater supplies.

Regulatory and technological measures must cover all major categories of point and non-point sources: domestic sewage; municipal and industrial landfills; mining operations; and agricultural practices. Overexploitation of aquifers leading to quality degradation must be prevented.

By the year 2000 assessments of known aquifers and their vulnerability to contamination should have commenced in all countries, while potential sources of groundwater pollution should have been identified and plans for their control developed.

4.6 Protection of aquatic ecosystems and living freshwater resources

Basis for action:

Aquatic living resources are threatened by a variety of human activities, particularly their overutilization with the objective of obtaining valuable products. Other human activities which produce threats to living freshwater resources include: accidental or incidental killing in the process of fishing for other species or in other ways; disturbance or harassment which may interfere with reproductive activities; and production of adverse environmental changes such as despoilment of breeding areas, reduction of food supplies and noxious chemical pollution.

An integrated approach to environmentally sustainable management of the water resources needs to include protection of aquatic ecosystems and living freshwater resources as a central goal. Living resources within aquatic ecosystems should be managed to optimize the benefits for human needs and the long-term sustainability of the ecosystem. These resources add to the value of water as an economic good and in return provide powerful justification for the protection of these ecosystems.

Strategy and programme targets:

The living resources within the aquatic ecosystem should be managed in such a way as to maximize the benefits for human needs derived from the ecosystem as a whole. Individual populations of commercially important species should be brought to highly productive levels on a sustainable basis. Improved conservation measures are required, based upon expanded research, for the depleted or threatened state of many living freshwater resources.

By the year 2000 all countries should put in place strategies for the environmentally sound management of their freshwater and associated coastal ecosystems; such strategies should consider fisheries, aquaculture, grazing, agricultural resources and biodiversity.

4.7 Targets and costs

(i) Targets

(a) All countries should have identified all potential sources of water supply and prepared outlines for their protection, conservation and rational use by the year 2000;
(b) All countries should have effective water pollution control programmes, defined as enforceable standards for major point-source discharges and high-risk non-point sources, commensurate with their socio-economic development by the year 2000;
(c) All countries should participate in inter-country and international water quality monitoring and management programmes such as Global Water Quality Monitoring Programme GEMS/WATER, UNEP's Environmentally Sound Management of Inland Waters, FAO's regional inland fishery bodies, and the RAMSAR Convention on Wetlands of International Importance Especially as Waterfowl Habitat by the year 2000;
(d) The prevalence of water-associated diseases should be drastically reduced by the year 2025, starting with the eradication of dracunculosis (Guinea worm) and onchocerciasis (river blindness) by the year 2000;
(e) All countries should have established biological, health, physical and chemical quality criteria for all water bodies (surface and ground water) with a view to the ongoing improvement of water quality by the year 2025.

(ii) Cost Estimates

Funds for pollution control have to be generated ultimately within each country or river basin through cost recovery and economic or fiscal instruments. The 'polluter pays principle' (PPP) has to be adopted in conformity with the notion of water as an economic good. Total costs, including those financed nationally, are estimated at about US$ 1.0 billion annually for the period 1993 to 2000. Of this amount about US$ 340 million would be needed annually, including about US$10 million for the strengthening of international institutions, from international sources on grant or concessional terms*.

The assessment of global environmental issues also includes water quality and aquatic ecosystems monitoring and assessment. River monitoring for global flux estimates has already been covered under water resources assessment, and the necessary funds have been indicated there.

* These costs are estimated for the establishment of legal, regulatory, institutional and technical measures to control water pollution.

5

Impacts of climate change on water resources

Climate change will mean increased water scarcity in many parts of the world Already critical water shortages in parts of semi-arid Africa are likely to become exacerbated. Credit: WMO.

5

Impacts of climate change on water resources

5.1 Background

In recent decades evidence has been building that human activities, particularly those resulting in the emission of 'greenhouse gases' into the atmosphere, are having an effect on climates globally. There is a growing realization, too, that climate changes will probably be evidenced most strongly through impacts on the hydrological cycle and through sea-level rise.

To address the climate change issues the Intergovernmental Panel on Climate Change (IPCC) was jointly set up by the World Meteorological Organization and the United Nations Environment Programme in 1988. The IPCC established three Working Groups to:

(a) Assess available scientific information on climate change;
(b) Assess environmental and socio-economic impacts of climate change;
(c) Formulate response strategies.

The main findings and recommendations of the IPCC Working Groups have been published in a series of reports (WMO/UNEP, 1990a, 1990b, 1990c, 1992) and later in book form (Bernthal, 1990, Houghton *et al.*, 1990 and Tegart *et al.*, 1990). The findings were discussed in detail at the Second World Climate Conference held in Geneva in 1990 (Jäger and Ferguson, 1991). As the potential impacts of climate change on water resources and sea-level rise are of very great significance to many countries, attention was also given to these issues at ICWE (WMO/UNESCO, 1991) and a special section in paragraphs 18.82 to 18.90 of Chapter 18 (Freshwater resources) of UNCED Agenda 21 (UN, 1993) addressed this subject.

The most significant findings from the IPCC documents regarding water resources and sea-level rise may be summarized as follows:

(a) While there is still considerable uncertainty, as a result of inadequate understanding of some key components of the Earth–atmosphere system, there is a broad consensus that global mean temperature is likely to increase during the next century at a rate of about 0.3 °C per decade. This rate might be altered as a result of radical changes in the emissions of greenhouse gases. It is likely that land surfaces would warm up more rapidly than the ocean and high northern latitudes warm more than the global mean in winter. There is likely to be considerable variability in climate change between regions and the confidence in predictions declines with areal detail. The ability to predict changes in precipitation and evaporation, key components in the hydrological cycle, is less than the ability to predict temperature changes. Natural variability is still large compared to likely change over periods up to a few decades and thus may mask global changes.

(b) The impacts on water resources are likely to be highly variable from region to region. Higher temperatures are likely to be associated with higher evaporation. Where combined with lower precipitation, this could lead to critically reduced water availability. On the other hand, some areas may experience higher precipitation with, perhaps, increased flood potential. Vegetation cover and agricultural productivity are likely to be affected through changes in evapo-transpiration rates, through changes in soil moisture and through changes in seasonal distribution of precipitation. Negative impacts on water resources will be most critical in regions which are already under water stress. For example, increasing temperatures and evaporation rates in the Sahel region of Africa could turn an already highly stressed area into a complete disaster zone.

(c) As a result of global temperature rise, sea-level is likely to rise at a rate of about 6 cm per decade, due to a combination of thermal expansion of the oceans coupled with release of ice from storage in mountain glaciers and ice sheets. Quite apart from loss of land, which would be critical for many small island nations, there would be impacts on groundwater and on estuaries through saline intrusion. Impacts on wetlands would be critical in some localities.

(d) The response strategies would be even more varied than the impacts, reflecting the diversity of socio-economic as well as physical conditions. Response strategies to water resource changes would be undertaken primarily at the national level; however, there would be an urgent need for international collaboration to address many of the issues.

5.2 Impact of climate change on freshwater resources, flooding and drought

Basis for action:

Among the most important impacts of climate change will be its effects on the hydrological cycle and water management systems, and through these on socio-economic systems. Higher temperatures and decreased precipitation would lead to decreased water supplies and increased water demands; they might also cause deterioration in the quality of freshwater bodies, putting strains on the already fragile balance between supply and demand in many countries. Even where precipitation might increase, there is no guarantee that this increase would occur at the time of year when it could be used. Increases in the incidence of extremes, such as floods and droughts, would cause increased frequency and severity of disasters.

Therefore, there is a need to acquire an adequate understanding of the potential impact of the predicted climate change on the availability and reliability of freshwater resources, on water demand, on the incidence of floods and droughts and on the consequences for the efficient management and safety of existing and future water-related projects and structures. This will permit the planning and implementation of effective counter-measures in the case of deleterious consequences and revised policies in the event of beneficial consequences.

Monitoring of climate change and its impact on freshwater bodies must be closely integrated with national and international programmes for monitoring the environment, in particular those concerned with the atmosphere and the hydrosphere. The analysis of data on climate change as a basis for developing remedial measures is a complex task. Extensive research is necessary in this area and due account has to be taken of the work of the Intergovernmental Panel on Climate Change, the World Climate Programme, the International Geosphere–Biosphere Programme (IGBP) and other relevant international programmes.

The development and implementation of response strategies require innovative use of technological means and engineering solutions, including the installation of flood and drought warning systems and the construction of new water resource development projects such as dams, aqueducts, well fields, waste–water treatment plants, desalination works, levees, banks and drainage channels. There is also a need for coordinated research networks to be promoted, such as those of the International Geosphere–Biosphere Programme/Global Change System for Analysis, Research and Training (IGBP/START) network.

Developmental work and innovation depend for their success on good staff training and motivation. International projects can help by enumerating alternatives, but each country needs to establish and implement the necessary policies and to develop its own expertise in the scientific and engineering challenges to be faced, as well as needing a body of dedicated individuals who are able to interpret the complex issues concerned for those required to make policy decisions. Such specialized personnel need to be trained, hired and retained in service, so that they may serve their countries in these tasks.

There is a need to build a capacity at the national level to develop, review and implement response strategies. Construction of major engineering works and installation of forecasting systems will require significant strengthening of the agencies responsible, whether in the public or the private sector. Most critical is the requirement for a socio-economic mechanism that can review predictions of the impact of climate change and possible response strategies and make the necessary judgements and decisions.

Table 23. *Impact of climate change on freshwater resources, flooding and drought*

Activities and related means of implementation	Level[1]	Considered by	
		ICWE[2]	UNCED[3]
1. Strengthen capabilities to collect, store and process water-related data, including data related to climate change. Monitor the hydrological regime, including soil moisture and groundwater balance, and monitor related climate factors, especially in the regions and countries most likely to suffer from the adverse effects of climate change and where the localities vulnerable to these effects should therefore be defined.	IN	X	X
2. Develop research programmes at the national level to contribute to regional and international research projects on the question of climate change, its early detection and its impact on the hydrological regime. These should address the situation in all countries and could involve case studies designed to develop and test specific methodologies for impact assessment.	IN	X	X
3. Develop and apply techniques and methodologies for assessing the potential adverse effects of climate change, through changes in temperature and precipitation on freshwater resources and the risk of floods and droughts.	IN	X	X
4. Assess the resulting social, economic and environmental impacts. Develop and initiate response strategies to counter the adverse effects, including changing groundwater levels and saline intrusion into aquifers. Develop agricultural activities based on brackish-water use.	IN	X	X
5. Promote cooperation between climatological and hydrological communities to develop predictions of climate change for individual seasons and for specific regions.	IN	X	

[1] Level of implementation: I=International; N=National; P=Provincial or sub-national; L=Local.
[2] Considered in ICWE *Report of the Conference* paragraphs 3.17–18.
[3] Considered in UNCED Agenda 21 paragraph: 18.82–90.

Strategy and programme targets:

The policy should be to acquire an adequate understanding of the threat of the impact of climate change on freshwater resources and on the incidence of floods and droughts so as to permit the planning and implementation of effective countermeasures. In particular, efforts should include:

(a) Long-term systematic observations of hydrological elements on global and regional scales;
(b) Studies of the role of hydrological processes in the cycling of energy, water and nutrients and of the impact of changes in the hydrological cycle on water resources and on their utilization;
(c) Full, open and timely exchange of relevant information;
(d) Examination of options for adapting water resource management practice to a changing climate.

Efforts should be directed through both national and international programmes involving comprehensive monitoring, research and policy review.

5.3 Impact of sea-level rise on water resources and flooding

Basis for action:

The final statement of the Second World Climate Conference (November, 1991) notes that global 'warming is expected to be accompanied by a sea-level rise of 65 cm ± 35 cm by the end of the next century'. It goes on to say that a 'sea-level rise would seriously threaten low-lying islands and coastal zones'.

A rise in sea-level would increase the threat of flooding from high tides, storm surges and river discharges in coastal and estuarine zones. In many instances this would make it too dangerous or uneconomic to maintain the current land use. It would also increase salt water intrusion in estuaries and coastal aquifers, the latter posing a major threat to what are often fragile freshwater supplies.

Strategy and programme targets:

Assessments should be made in all low-lying islands and coastal zones of the potential impact of any rise in sea-level. This should include assessments on the quality and quantity of freshwater resources and on the risk of flooding. Both national and international bodies have roles to play in this respect.

Table 24. *Impact of sea-level rise on water resources and flooding*

Activities and related means of implementation	Level[1]	Considered by	
		ICWE[2]	UNCED[3]
1. Strengthen capabilities to monitor rises in sea-level and related hydrological parameters.	IN	X	X
2. Develop techniques for assessing the potential impact of sea-level rise on freshwater resources and flood risk.	IN	X	X
3. Assess the likely socio-economic and environmental impacts of a rise in sea-level. Develop and implement response strategies.	IN	X	X

[1] Level of implementation: I=International; N=National; P=Provincial or sub-national; L=Local.
[2] Considered in ICWE *Report of the Conference* paragraphs 3.19–20.
[3] Considered in UNCED Agenda 21 paragraph: 18.82–90.

5.4 Targets and costs

(i) Targets

(a) To understand and quantify the threat of the impact of climate change on freshwater resources by the year 2000;
(b) To facilitate the implementation of effective national counter-measures.

(ii) Cost estimates

It is estimated that over the period 1993 to 2000 the average total annual cost of implementing the activities of this programme will be about US$100 million, including about US$ 40 million from the international community on grant or concessional terms*.

* The impacts of climate change on freshwaters are costed with regard to sea-level rise, droughts and increased flood risk.

6

Water and sustainable urban development

Water availability in acceptable quality is a critical problem in many cities in the developing world. This scene from Calcutta typifies the problems for many urban inhabitants. Credit. UN.

6

Water and sustainable urban development

6.1 Background

Urbanization and industrialization linked with rapid population growth have been major engines for national economic growth. These tendencies are expected to continue and, in some countries, to accelerate. Urban domestic and industrial consumers are using even larger shares of available water resources and are, at the same time, degrading these resources with their wastes. Urgent actions are required to improve the effectiveness of use of water resources, if their contribution to human well-being and productivity is to be sustained. Urban water resources management is treated in detail by Gupta (1992) and Rogers (1992).

Early in the next century, more than half of the world's population will be living in urban areas. By the year 2025, that proportion will have risen to 60 %, comprising some 5 billion people. Rapid urban population growth and industrialization are putting severe strains on the water resources and environmental protection capabilities of many cities. Special attention needs to be given to the growing effects of urbanization on water demands and usage and to the critical role played by local and municipal authorities in managing the supply, use and overall treatment of water, particularly in developing countries, for which special support is needed. Scarcity of freshwater resources and the escalating costs of developing new resources have a considerable impact on all forms of national development and economic growth (industrial, agricultural and human settlement). Better management of urban water resources, including the elimination of unsustainable consumption patterns, can make a substantial contribution to the alleviation of poverty and improvement of the health and quality of life of the urban and rural poor. A high proportion of large urban agglomerations is located around estuaries and in coastal zones. Such an arrangement leads to pollution from municipal and industrial discharges, which, combined with overexploitation of available water resources threatens the marine environment as well as the supply of freshwater resources.

A major development objective is to support the efforts and capacities of local and central governments in their task of sustaining national development and productivity through environmentally sound management of water resources for urban use. Essential for the support of this objective are the identification and implementation of strategies and actions to ensure the continued supply of affordable water for present and future needs and the reversal of current trends of resource degradation and depletion.

ICWE endorsed three key strategic principles for setting priorities and choice of action programmes for the use and management of water resources:

- Water should be considered an economic good having a value consistent with its most valuable potential use;
- The above principle necessitates new institutional approaches to the management of water resources. Capacity building, especially institutional development, should therefore receive priority attention;
- In the choice of sectoral priorities for action programmes, relatively more attention should be given to the management of wastes (reduction, reuse and recycling, collection, treatment and disposal).

To prevent exaggerated expectations about the rate of progress which is possible in resolving important issues in implementing the above principles, action programmes should be consistent with realistic appraisal of the resources available to implement them, while at the same time the international community should mobilize resources to support the proposed actions.

The 1980s saw considerable progress in the development and application of low-cost water supply and sanitation technologies. The programme envisages continuation of this work, with particular emphasis on development of appropriate sanitation and waste disposal technologies for low-income high-density urban settlements. There should also be international information exchange, to ensure a widespread recognition among planners and designers of the availability and benefits of appropriate low-cost technologies. Public awareness campaigns should also include components designed to overcome user resistance to such appropriate low-cost technologies by emphasizing the benefits of reliability and sustainability.

Implicit in virtually all elements of this programme is the need for progressive enhancement of the training and career development of personnel at all levels in sector institutions. Specific programme activities will involve the training and retention of staff with skills in community involvement, low-cost technology, financial management, and integrated planning of urban water resources management. Special provision should be made for mobilizing and facilitating the active participation of women, youth, indigenous people and local communities in water management teams and for supporting the development of water associations and water committees, with appropriate training of such personnel as treasurers, secretaries and caretakers. Special education and training programmes for women should be launched on the protection of water resources and water quality within urban areas.

In combination with human resource development, strengthening of institutional, legislative and management structures are key elements of the programme. A prerequisite for progress in enhancing access to water and sanitation services is the establishment of an institutional framework that ensures that the real needs and potential contributions of currently unserved populations are reflected in urban development planning. The multisectoral approach, which is a vital part of urban water resources management, requires institutional linkages at the national and city levels, and the programme includes proposals for establishing intersectoral planning groups. Proposals for greater pollution control and prevention depend for their success on the right combination of economic and regulatory mechanisms, backed by adequate monitoring and surveillance and supported by enhanced capacity to address environmental issues on the part of local Governments.

Establishment of a set of appropriate design standards, water quality objectives and discharge consents is therefore among the proposed activities. The programme also includes support for strengthening the capability of water and sewerage agencies and for developing their autonomy and financial viability. Operation and maintenance of existing water and sanitation facilities have been recognized as having serious shortcomings in many countries. Technical and financial support are needed to help countries correct present inadequacies and build up the capacity to operate and maintain rehabilitated and new systems.

6.2 Efficient and equitable allocation of water resources

Basis for action

At present in developing countries, estimates indicate that about 85% of available water resources are used for agriculture, 10% for industry, and 5% for domestic supplies. In many countries, there are serious inefficiencies both between sectors and within sectors. Increased efficiency in water utilization for agriculture should be promoted in order to relieve pressure on resources and to prevent water availability becoming a limiting factor in industrial growth.

As demand grows and resources diminish, the allocation of scarce resources is becoming a major political issue and priorities have to be established which balance health improvement, desires for food security, environmental protection and economic growth. The economic, social and environmental priorities to be established should take into account the availability and long-term sustainability of water resources ensuring, as a top priority, the availability of sufficient, affordable domestic supplies, in order to meet the basic needs of the community and especially the needs of the very poor, particularly during droughts.

Table 25. *Efficient and equitable allocation of water resources*

Activities and related means of implementation	Level[1]	Considered by	
		ICWE[2]	UNCED[3]
1. Inform public opinion and encourage Governments to develop priorities according to economic criteria for the allocation of water resources to achieve long-term sustainability. Reconcile city development planning with the availability and sustainability of water resources.	NP	X	X
2. Strengthen institutional capacity, especially at the local level, and provide technical support for the introduction and application of water charges and pollution penalties which reflect the marginal and opportunity costs of water, especially for productive activities.	INPL	X	X
3. Foster water conservation, recycling and pollution reduction through all available means, including economic and regulatory incentives and technical devices.	NPL	X	
4. Provide technical and financial support at all levels for assessment and monitoring to help safeguard the availability and quality of surface and groundwater resources and reduce pollution load in all sectors.	IN	X	
5. Develop standard techniques that present in a form understandable by non-experts the ranking of alternative resource allocations according to multiple criteria and alternative assumptions about the future.	N		
6. Collect and compile data on water use and polluting load in all sectors, with projections for future water use and polluting load based on alternative conservation/reuse strategies.	NP		
7. Monitor compliance with abstraction licences and discharge consents, and ensure effective collection of charges and imposition of penalties.	NPL		

[1] Level of implementation: I=International; N=National; P=Provincial or sub-national; L=Local.
[2] Considered in ICWE *Report of the Conference* section 5.
[3] Considered in UNCED Agenda 21 paragraph: 18.56–64.

Strategy and programme targets:

Satisfaction of the basic water needs of the urban population especially the urban poor:

- Establish economic, social and environmental priorities which take account of the availability and long-term sustainability of water resources, and allocate water resources accordingly, ensuring as a top priority the availability of sufficient affordable domestic supplies.
- Introduce charging structures which reflect the full marginal costs (including opportunity costs) for all water supplies, with subsidies restricted to meeting the basic needs of the very poor.
- By the year 2000, have in place a water resources master plan which matches development objectives with water resource sustainability, and which enables cities to plan on the basis of assured water allocations.

6.3 Protection against depletion and degradation of water resources

Basis for action:

Pollution from untreated municipal and industrial wastes is causing health-threatening conditions in surface water resources. At the same time, over-abstraction and contamination are depleting groundwater resources. The costs of providing new water are rising rapidly, while protection, conservation and reuse could enable demands to be met much more economically.

Experience in some developed countries demonstrates great scope for successive use and reuse of water for both industry and agriculture. To achieve economic gains, resources need to be protected, and water charges and pollution penalties have to reflect the true value of water.

Strategy and programme targets:

As part of an overall strategy to protect health and the environment and to make the most economic use of all available water resources, present pollution trends should be reversed to improve water quality progressively. In this context, development funding agencies should take the initiative to encourage the incorporation of a significant element of environmental improvement in water-related projects.

Within 10 years, programmes to provide sanitary containment or treatment for at least 50% of the pollution load (in terms of biological oxygen demand) from domestic wastes should be initiated. By 2015 every country should have achieved river water quality (varying from location to location) which safeguards supplies for downstream users.

Table 26. *Protection against depletion and degradation of water resources*

Activities and related means of implementation	Level[1]	Considered by	
		ICWE[2]	UNCED[3]
1. Use scientifically established guidelines and set objectives for the protection of all river systems and groundwater resources; translate the objectives into discharge quality and reuse standards for upstream municipal and industrial effluents; implement monitoring programmes, supported by enforced legislation and pricing mechanisms.	INPL	X	
2. Implement national programmes to introduce sanitary waste disposal facilities, based on environmentally sound low-cost and upgradable technologies, and to ensure that investment in public water supply is accompanied with appropriate investment in the removal, recycling, safe reuse and disposal of municipal wastes and surface water drainage.	NPL	X	X
3. Implement urban storm-water runoff and drainage programmes.	NPL		X
4. Encourage the best management practices for the use of agrochemicals with a view to minimizing their impact on water quality.	NPL	X	X
5. Include water consumption and effluent load among the criteria conditioning the choice and approval of the location of new industrial sites or expansions thereof and establish standards for effluent quality for existing and new industries.	NP	X	X
6. Protect watersheds from depletion and degradation of their forest cover and from harmful upstream activities. Promote research into the contribution of forests to sustainable water resources development.	NPL	X	X
7. Increase environmental awareness through education and public relations campaigns to stimulate behavioural change to conserve water, combat pollution, and increase disaster preparedness.	INPL	X	X
8. Enhance and promote cross-sectoral information exchange and applied research on improved recycling techniques, groundwater treatment and protection methods and surface water drainage to abate runoff pollution.	INP	X	X

[1] Level of implementation: I=International; N=National; P=Provincial or sub-national; L=Local.
[2] Considered in ICWE *Report of the Conference* section 5.
[3] Considered in UNCED Agenda 21 paragraph: 18.56–64.

Table 27. *Institutional/legal/management reforms*

Activities and related means of implementation	Level[1]	Considered by	
		ICWE[2]	UNCED[3]
1. Establish institutional and legislative frameworks for water management and pollution protection at national and local levels, adopting a city-wide approach to the management of water resources, and integrating water resources planning and land-use management; develop and apply regulatory and economic instruments; and undertake monitoring and surveillance. Develop legislation and policies to promote investments in urban water and waste management, reflecting the major contribution of cities to national economic development. Provide seed money and technical support to the local handling of materials supply and services.	INPL	X	X
2. Develop institutional frameworks which bring together water utilities, non-governmental organizations, the private sector and community groups to exchange views, contribute skills and take decisions on water supply and sanitation projects. Support intersectoral planning involving relevant sectoral agencies at all administrative levels.	NPL	X	X
3. Encourage the autonomy and financial viability of city water, solid waste and sewerage utilities.	NPL		X
4. Train staff at all levels with skills in community involvement, low-cost technologies, and financial management and for undertaking hygiene education programmes, with a focus on women and children. Create and maintain a cadre of professionals and semi-professionals, for water, waste-water and solid waste management.	NPL	X	X
5. Undertake international collaboration and information exchange in support of institutional reforms.	I	X	

[1] Level of implementation: I=International; N=National; P=Provincial or sub-national; L=Local.
[2] Considered in ICWE *Report of the Conference* section 5.
[3] Considered in UNCED Agenda 21 paragraph: 18.56–64.

Table 28. *Resource mobilization*

Activities and related means of implementation	Level[1]	Considered by	
		ICWE[2]	UNCED[3]
1. Undertake a global programme to promote public information and communication campaigns to mobilize support for achieving sustainable urban development, emphasizing the extent to which this is threatened by current trends.	INPL	X	
2. Provide technical support and building capacities: (a) to ensure financial viability; (b) to encourage community involvement, including the participation of women, in decision-making including investments; (c) to reduce unaccounted-for water and to take full advantage of recycling opportunities in municipal wastewater and solid waste disposal.	INPL	X	
3. Undertake applied research to extend economic evaluation techniques to include environmental considerations in full.	INPL	X	

[1] Level of implementation: I=International; N=National; P=Provincial or sub-national; L=Local.
[2] Considered in ICWE *Report of the Conference* section 5.
[3] Considered in UNCED Agenda 21 paragraph: 18.56–64.

6.4 Institutional/legal/management reforms

Basis for action:

As scarcity increases, water resources have a greater impact on development planning. Urban water utilities are often underfunded, understaffed and working in an institutional vacuum.

Urban water and sanitation services need to be integrated with land use, housing and environmental protection. Urban water resource management should, therefore, be integrated with all aspects of national and regional planning processes affecting the region in which the city is sited.

A prerequisite for effective water and waste management is an institutional and legislative framework able to resolve conflicting demands and enforce standards. While there should be overall integration of services, water and sanitation agencies need sufficient autonomy to control their own finances, to respond to local needs, and to attract and retain the right staff.

Strategy and programme targets:

Water resources should be managed at the lowest appropriate level by institutions capable of managing water distribution and demand.

Financial viability and autonomy of urban water and sewerage agencies should be encouraged in some cases through privatization.

By the year 2000 all countries should have arrangements including the necessary monitoring for enforcing water and effluent standards which should reflect the 'polluter pays' principle.

6.5 Resource mobilization

Basis for action:

Reversal of the present trends in urban degradation is going to require much higher levels of investment by municipal authorities, governments and external support agencies. The economic case for greater investment is powerful: unless the twin problems of water scarcity and pollution are tackled urgently and effectively, shortage of water will become an overriding constraint on national economic growth in a number of countries.

This message must be made apparent to political leaders and development support agencies, as quickly and as frequently as possible.

Strategy and programme targets:

Make investments in urban water and waste management commensurate with the major contribution of cities to national economic growth.

Convert the cash paid to water vendors by currently unserved populations into investment in reliable public supplies.

Employ tariffs and collection systems which assure adequate cost recovery for the upkeep and extension of services.

Develop fully costed estimates and identify funding sources for achieving coverage targets and pollution protection objectives.

6.6 Targets and costs

(i) Targets

(a) To ensure that all urban residents have reasonable access to at least 40 litres per capita per day of safe water and that 75% of the urban population is provided with on-site or community facilities for sanitation by the year 2000;

(b) To establish and apply quantitative and qualitative discharge standards for municipal and industrial effluents by the year 2000;

(c) To ensure that 75% of the solid wastes generated in urban areas is collected and recycled or disposed of in an environmentally safe way by the year 2000.

(ii) Cost Estimates

It is estimated that the average total annual cost (1993–2000) of implementing the activities of this programme would be about $20 billion, including about $4.5 billion from the international community on grant or concessional terms*.

* Cost estimates for drinking water supply and basic sanitation in rural areas are treated elsewhere.

7

Water for sustainable food production and rural development

Water for irrigation is essential for agriculture and in many countries women constitute a very high percentage of the labour force in the agricultural field. Credit: 'Women, water supply and sanitation', ILO, UN-INSTRAW, UN-DTCD.

7
Water for sustainable food production and rural development

7.1 Background

The population of the globe is expected to reach 8 billion by the year 2020. The consequent dramatic growth in food demand and the pressures upon the natural resource base pose an enormous challenge to the rural sector. The latter will not only have to increase food production substantially, but being the major user of water resources, it will have to release freshwater resources to meet increasing demands for domestic, industrial and ecosystem management purposes. At the same time the rural sector will have to contribute significantly to the conservation of the natural resource base and especially to the husbandry of the aquatic environment.

For many developing countries, agriculture remains the dominant economic sector and the majority of their population is rural. While the rural population will grow in absolute terms, it will not only have to meet its own rising food requirements but additionally it will have to feed a rapidly increasing urban population. Meeting the basic needs of the rural populations for drinking water, food, shelter, clothing, sanitation and primary health care, and the environmentally sustainable transformation of the subsistence agriculture to a productive and economically viable venture form the basis of rural development.

Sustainability of food production increasingly depends on sound and efficient water use and conservation practices (cf. Rydzewski and Abdullah, 1992) consisting primarily of irrigation development and management and including water management with respect to rain-fed areas, livestock water supply, inland fisheries and agro-forestry. Achieving food security is a high priority in many countries, and agriculture must not only provide food for the rising population, but also save water for other uses. The challenge is to develop and apply water-saving technology and management methods and enable communities to introduce institutions and incentives for the rural population to adopt new approaches, for both rain-fed and irrigated agriculture. The rural population must also have better access to a potable water supply and to sanitation services.

It is an immense task but not an impossible one, provided appropriate policies and programmes are adopted at all levels–local, national and international. While significant expansion of the area under rain-fed agriculture has been achieved during the past decade, the productivity response and sustainability of irrigation systems have been constrained by problems of waterlogging and salinization. Financial and market constraints are also a common problem. Soil erosion, mismanagement and overexploitation of natural resources and acute competition for water have all influenced the extent of poverty, hunger and famine in the developing countries. Soil erosion due to overgrazing by livestock is also often responsible for the siltation of lakes. Most often, the development of irrigation schemes is supported neither by environmental impact assessments, identifying hydrological consequences of inter-basin transfers within watersheds, nor by the assessment of social impacts on peoples in river valleys.

The non-availability of water supplies of suitable quality is a significant limiting factor to livestock production in many countries, and improper disposal of animal wastes can, in certain circumstances, result in pollution of water supplies for both humans and animals. The drinking water requirements of livestock vary according to species and the environment in which they are kept. It is estimated that the current global livestock drinking water requirement is about 60 billion litres per day. Based on livestock population growth estimates, this daily requirement is predicted to increase annually by 0.4 billion litres per day in the foreseeable future.

Freshwater fisheries in lakes and streams are an important source of food and protein. Inland water fisheries should be so managed as to maximize the yield of aquatic food organisms in an environmentally sound manner. This requires the conservation of water quality and quantity, as well as of the functional morphology of the aquatic environment. On the other hand, fishing and aquaculture may themselves damage the aquatic ecosystem; hence their development should conform to guidelines for impact limitation. Present levels of production from inland fisheries, from both fresh and brackish water, are about 7 million tonnes per year and could increase to 16 million tonnes per year by the year 2000; however, any increase in environmental stress could jeopardize this rise.

An International Action Programme on Water and Sustainable Agricultural Development (IAP-WASAD) has been initiated by FAO in cooperation with other international organizations (FAO, 1990). The main objective of the Action Programme is to assist developing countries in planning, developing and managing water resources on an integrated basis to meet present and future needs for agricultural production, taking into account environmental considerations.

The Action Programme has developed a framework for sustainable water use in the agricultural sector and identified priority areas for action at national, regional and global levels. Quantitative targets for new irrigation development, improvement of existing irrigation schemes and reclamation of waterlogged and salinized lands through drainage for 130 developing countries are estimated on the basis of food requirements, agro-climatic zones and availability of water and land.

The development of new irrigation areas at the above-mentioned level may give rise to environmental concerns in so far as it implies the destruction of wetlands, water pollution, increased sedimentation and a reduction in biodiversity. Therefore, new irrigation schemes should be accompanied by an environmental impact assessment, the scope and extent of which would depend on the scale of the project and the potential for negative impacts. When considering proposals for new irrigation schemes, consideration should also be given to a more rational exploitation, and an increase in the efficiency or productivity, of any existing schemes capable of serving the same localities. Technologies for new irrigation schemes should be thoroughly evaluated, including their potential conflicts with other land uses. The active involvement of water-user groups is a supporting objective.

It should be ensured that rural communities of all countries will have access to safe water in sufficient quantities and adequate sanitation to meet their health needs and maintain the essential qualities of their local environments (cf. WHO, 1991).

The objectives with regard to water management for inland fisheries and aquaculture include conservation of water-quality and water-quantity requirements for optimum production and prevention of water pollution by aquacultural activities. The Action Programme seeks to assist member countries in managing the fisheries of inland waters through the promotion of sustainable management of capture fisheries as well as the development of environmentally sound approaches to intensification of aquaculture.

The objectives with regard to water management for livestock supply are twofold: provision of adequate amounts of drinking water and safeguarding of drinking water quality in accordance with the specific needs of different animal species. This entails maximum salinity levels and the absence of pathogenic organisms. No global targets can be set owing to large regional and intra-country variations.

There is an urgent need for countries to monitor water resources including water quality, water and land use, and crop production; compile inventories of type and extent of agricultural water development and of present and future contributions to sustainable agricultural development; evaluate the potential for fisheries and aquaculture development; and improve the availability and dissemination of data to planners, project designers, technicians, farmers and fishermen. Priority requirements for research are as follows:

Table 29. *Efficient and rational allocation of water*

Activities and related means of implementation	Level[1]	Considered by	
		ICWE[2]	UNCED[3]
1. Provide technical and financial support for assessment of the water resources and their allocation among users through: (a) hydrological studies, establishment of monitoring stations, and analysis of data to establish water resources profiles; (b) compilation of data on water use by sector, with projections for future water use; (c) promotion of political judgment on development priorities and corresponding allocation of water resources; (d) development of river basin and transboundary management plans to safeguard future supplies.	INPL	X	X
2. Develop policies and programmes to manage scarce water resources for agriculture through: (a) development of long-term strategies and practical implementation programmes for agricultural water use under scarcity conditions with competing demands for water; (b) formulation of specialized programmes focused on drought preparedness, with emphasis on food scarcity and environmental safeguards; (c) promotion and enhancement of waste-water reuse in agriculture.	INPL	X	X
3. Promote and actively support water conservation and water-use efficiency in agriculture through: (a) ensuring that the technical, institutional and budgetary requirements for future operation and maintenance are provided for in the planning, design and implementation phases of new projects; (b) evaluation of the scope for rehabilitating existing malfunctioning systems as an alternative to investing in new projects; (c) establishing preventive maintenance schedules, leak detection programmes, and regular quality surveillance; (d) introducing and actively using baseline assessments of readily measurable indicators, effective monitoring mechanisms during project implementation and evaluation of achievements against set objectives; (e) maximizing water conservation in new irrigation schemes and improving drainage in wet and saline croplands; (f) increasing water-use efficiencies in rain-fed agriculture through measures including flood management and drought mitigation; (g) enhancing the above actions on water management by complementary measures, such as widespread introduction of drought resistant crops, insect and rodent control, effective storage and transportation.	NPL	X	X
4. Strengthen institutional and technical support for the introduction and application of water charges and pollution penalties which reflect the marginal and opportunity cost of water by: (a) recognizing water as a social, economic and strategic good in irrigation planning and management; (b) developing and implementing pricing mechanisms which encourage conservation and protection of water resources; (c) making available affordable supplies for meeting the basic needs of the unserved poor; (d) monitoring compliance with abstraction licences and discharge consents, and ensuring effective collection of charges and imposition of penalties.	INPL	X	X

[1] Level of implementation: I=International; N=National; P=Provincial or sub-national; L=Local.
[2] Considered in ICWE *Report of the Conference* section 6.
[3] Considered in UNCED Agenda 21 paragraph: 18.65–81.

(a) Identification of critical areas for water-related adaptive research;
(b) Strengthening of the adaptive research capacities of institutions in developing countries;
(c) Enhancement of the translation of research results on water-related farming and fishing systems into practical and accessible technologies and provision of the support needed for their rapid adoption at the field level.

Transfer of technology, both horizontal and vertical, needs to be strengthened. Mechanisms to provide credit, input supplies, markets, appropriate pricing and transportation must be developed jointly by countries and by external support agencies. Integrated rural water supply infrastructure, including facilities for water-related education and training and support services for agriculture, should be expanded for multiple uses and should assist in developing the rural economy. In summary, the key strategic principles for holistic and integrated environmentally sound management of water resources in the rural context are:

(a) Water should be regarded as a finite resource having an economic value with significant social and economic implications reflecting the importance of meeting basic needs;
(b) Local communities must participate in all phases of water management, ensuring the full involvement of women in view of their crucial role in the practical day-to-day supply, management and use of water;
(c) Water resource management must be developed within a comprehensive set of policies for (i) human health; (ii) food production, preservation and distribution; (iii) disaster mitigation plans; (iv) environmental protection and conservation of the natural resource base;
(d) It is necessary to recognize and actively support the role of rural populations, with particular emphasis on women.

7.2 Efficient and rational allocation of water

Basis for action:

In the future, a major part of the available freshwater will be taken up to supplement rainfall for agricultural production. Such a use of scarce water resources will arise because of the water required to satisfy the needs for food, fuel, fodder, fibre and timber of rapidly growing populations in developing economies. At the same time, the demands for good-quality water for drinking and sanitation and for industrial use are rising sharply, especially in the rapidly growing urban areas. Safeguarding good quality water is moreover necessary for fisheries and aquaculture, the maintenance of valuable natural ecosystems and for environmental protection in general.

This combination of increasing demands on finite freshwater resources makes them ever scarcer. More efficient use of resources, especially in the agricultural sector, and rational allocation between the various demand sectors are called for. The main strategies should ensure that water users realize the scarcity value of the resource and the incentives to promote its preservation must be established. Measures would include demand management in the form of charging systems designed to promote efficient and just use of water; cost-recovery policies to provide secure, sustained, efficient operation and maintenance of water supply systems; education and public information programmes; and legal entitlements for access to water resources. Such measures will have to be introduced with due consideration for the cultural, social and ecological values of water in the community concerned. Simultaneously, priority should be given to meeting the basic needs of the poor, including drinking water and small-scale agriculture. Prerequisites to resolving the competing demands are: comprehensive resource inventory and evaluation of existing land and water needs; the promotion of water storage and saving devices; and sound water use at river basin and village levels.

The quality of freshwater is declining in many parts of the world due to human induced land degradation, salinization, and pollution by toxic compounds and domestic, industrial and agricultural contaminants, including farmyard manure. The main strategy to combat such degradation is to arrest the problem at its source, through incentives and regulations for soil and water conservation measures that are environmentally sound . Close monitoring of all waste disposal and contamination is required as well as application of appropriate legal and administrative controls and the establishment of requirements for polluters to cover the cost of recovery of the water quality. To prevent losses in quantity and quality of agricultural produce and to protect human health, water quality standards for agricultural, drinking and sanitation uses should be set and appropriate mechanisms put in place for their effective implementation.

At present in many developing countries, it is estimated that about 85% of available water resources are used for agriculture, 10% for industry, and 5% for domestic supplies. This distribution of water resources will need to be reevaluated, especially where water resources are scarce.

As demand grows and resources diminish, the allocation of scarce resources is becoming a major political issue. Priorities have to be established which balance desires for food security, health improvement, social development, environmental protection, and – as a prerequisite for all those things – economic growth.

Flood recession agriculture is a primary source of food supply on many major flood plains. Its value in meeting the basic needs of the rural poor should therefore be given special recognition when considering management of basin resources. Due consideration should be given to maintaining minimum river flows downstream when designing water-management schemes.

In order to meet the basic needs of the rural populations, water development and management will have to be considered in an integrated manner. This integrated approach has to consider sustainable development programmes, including institutional and human resources development, protection of the environment and preservation of food supplies.

Strategy and programme targets:

- Establish economic, social and environmental priorities which take account of the availability and long-term sustainability of water resources, and allocate water resources accordingly.
- Introduce charging structures which reflect the full marginal costs (including opportunity costs) for all water supplies.
- Develop a water resources master plan which matches development objectives with water resource sustainability, and which enables rural communities and villages to plan on the basis of assured water allocations.

7.3 Protection against depletion and degradation of water resources

Basis for action:

Under many irrigated conditions, crop yields and quality are severely affected by high salinity or specific ion toxicity of water. Equally, improper agricultural activities and discharges of wastewater from rural settlements have led to pollution of surface- and groundwaters and render these sources unsuitable for potable use.

Strategy and programme targets:

- Assessing the quality of water for irrigation and drinking purposes;
- Developing agricultural water-use and agronomic practices which minimize water pollution;
- Implementing proper treatment and use of wastewater in agriculture;
- Identifying and monitoring sources of drinking water in rural areas.

7.4 Efficient use of water at the scheme and farm level

Basis for action:

In many parts of the world, there are deficiencies of design and of operation both at the irrigation scheme level and at the farm level. These can result in waterlogging and salinization of irrigated lands, avoidable water losses, habitats favourable to water-related diseases, and environmental degradation.

Table 30. *Protection against depletion and degradation of water resources*

Activities and related means of implementation	Level[1]	Considered by	
		ICWE[2]	UNCED[3]
1. Establish and improve biological, physical and chemical water quality criteria for agricultural water-users and for marine and riverine ecosystems.	NP	X	X
2. Implement monitoring programmes, backed by legislation and pricing mechanisms, to monitor water quality and to control polluting discharges.	NPL	X	X
3. Evaluate and monitor the quality of water for agricultural use and adopt appropriate management practices.	INPL	X	X
4. Demonstrate the need for improved agricultural and forestry practices to prevent the degradation and depletion of water resources for downstream users. Prevent adverse effects of agricultural activities on water quality for other social and economic activities and on wetlands, *inter alia*, through optimal use of on-farm input and the minimization of the use of external input in agricultural activities.	IN	X	X
5. Adopt appropriate water management and agronomic practices to prevent agricultural water pollution. Minimize soil erosion and sedimentation.	INPL	X	X
6. Encourage proper treatment and use of municipal sewage and farm manure and their eventual safe disposal.	INPL	X	X
7. Minimize use of agrochemicals by practicing integrated nutrient and pest management practices.	INPL	X	X
8. Increase environmental awareness and stimulate behavioural change to conserve water and combat gross pollution of water resources. Educate communities about the pollution-related impacts of the use of fertilizers and chemicals on water-quality, food safety and human health.	NPL	X	X

[1] Level of implementation: I=International; N=National; P=Provincial or sub-national; L=Local.
[2] Considered in ICWE *Report of the Conference* section 6.
[3] Considered in UNCED Agenda 21 paragraph: 18.65–81.

Table 31. *Efficient use of water at the scheme and farm level*

Activities and related means of implementation	Level[1]	Considered by	
		ICWE[2]	UNCED[3]
1. Improve irrigation infrastructure and introduce improved irrigation and agronomic practice in order to increase the efficiency and productivity in agricultural water use for better utilization of limited water resources.	IN L	X	X
2. Strengthen extension services and water and soil management adaptive research under irrigation and rainfed conditions.	INPL	X	X
3. Monitor and evaluate irrigation project performance to ensure, *inter alia*, the optimal utilization and proper maintenance of the project.	PL	X	X
4. Support water-users groups with a view to improving management performance at the local level through: (a) provision of adequate technical advice and support and enhancement of institutional collaboration; (b) promotion of local initiatives for the integrated development and management of water resources; (c) promotion of a farming approach for land and water management that takes account of the level of education and the capacity for mobilization of local communities.	NPL	X	X
5. Encourage water pricing and cost-recovery mechanisms.	INPL	X	
6. Support the appropriate use of brackish water for irrigation.	I L	X	X
7. Formulate and develop long-term irrigation and water-supply programmes, at the appropriate scale, for humans and livestock and for water and soil conservation, taking into account their effects on the local level, the economy and the environment. Plan and develop multipurpose hydroelectric power schemes, making sure that environmental concerns are duly taken into account.	NPL		X

[1] Level of implementation: I=International; N=National; P=Provincial or sub-national; L=Local.
[2] Considered in ICWE *Report of the Conference* section 6.
[3] Considered in UNCED Agenda 21 paragraph: 18.65–81.

Strategy and programme targets:

- Improvement of irrigation and on-farm infrastructures;
- Introduction of cost-recovery mechanisms;
- Demand management;
- Close monitoring of irrigation system performance and water management at the field level;
- Promotion of adaptive technological research and development as well as the dissemination of results;
- Strengthening of irrigation institutions, including water-users associations.
- Improving and modernizing the systems on 12 million ha of existing irrigated lands by the year 2000.

7.5 Small-scale water programmes

Basis for action:

Improvement of rainfed agriculture by small-scale water programmes including collective well-irrigation systems, small reservoir or tank irrigation schemes, multi-purpose water harvesting projects, village drinking water supply and community garden programmes can fulfil many rural community needs and can be sustainable. Such small-scale programmes may include diverting and storing of temporary excesses of rain waters and their subsequent use.

Strategy and programme targets:

The primary strategy consists of providing incentives and technical and institutional support to local communities to develop and manage water resources to meet their multiple needs. By the year 2000 about 10 million ha of rainfed arable lands should be improved through small-scale water programmes.

7.6 Waterlogging, salinity control and drainage

Basis for action:

Flooding, lack of adequate drainage, poor operation and maintenance of irrigation schemes, and inefficient water application at the farm level have contributed to surface water stagnation, waterlogging and salinization of many irrigated lands and low-lying areas. They have resulted in not only loss in productivity of agricultural lands but also deterioration of the environment.

Table 32. *Small-scale water programmes*

Activities and related means of implementation	Level[1]	Considered by	
		ICWE[2]	UNCED[3]
1. Develop methods to assess environmental impacts of all small-scale water programmes.	IN	X	
2. Promote integrated farming approaches in accordance with educational levels and local capacities.	I L	X	
3. Enhance community participation and ensure the role of women in community-based water projects.	IN L	X	
4. Promote environmental protection through use of appropriate, low-cost technologies with low waste generation.	N L	X	
5. Promote safe disposal/use of human excreta/waste through appropriate sanitation and hygiene education.	IN L	X	
6. Introduce water harvesting techniques in rainfed arable lands.	I L	X	
7. Provide technical and investment support for small-scale irrigation, water supply, sanitation and conservation projects.	INPL	X	
8. Target interventions to optimize investments and increase productivity.	IN	X	
9. Educate and encourage local groups to adopt environmentally sound small-scale water programmes.	IN L	X	

[1] Level of implementation: I=International; N=National; P=Provincial or sub-national; L=Local.
[2] Considered in ICWE *Report of the Conference* section 6.
[3] Considered in UNCED Agenda 21 paragraph: 18.65–81.

Strategy and programme targets:

The primary strategy consists of the reduction of the sources of excess water and provision of artificial drainage in existing wet croplands when necessary. Artificial drainage should be provided to about 7 million ha of irrigated lands during the period 1993–2000. It should be reinforced by the introduction of efficient pricing and cost-recovery mechanisms. Flood and rain waters causing temporary ponding on extensive plains can be contained for subsequent dry-season crop growth.

Table 33. *Waterlogging, salinity control and drainage*

Activities and related means of implementation	Level[1]	Considered by	
		ICWE[2]	UNCED[3]
1. Promote and introduce surface drainage of low-lying rainfed agriculture areas to prevent flooding by rainfall and improper management practices.	NPL	X	X
2. Introduce artificial drainage to control groundwater build-up and salinization of farm lands.	I PL	X	X
3. Promote conjunctive use of groundwater and surface water and monitor the water balance at basin and project levels.	I PL	X	X
4. Ensure proper installation, operation and maintenance of drainage systems in irrigated areas in arid and semi-arid regions including reuse or safe disposal of the drainage water.	INPL	X	X

[1] Level of implementation: I=International; N=National; P=Provincial or sub-national; L=Local.
[2] Considered in ICWE *Report of the Conference* section 6.
[3] Considered in UNCED Agenda 21 paragraph: 18.65–81.

7.7 Water for livestock

Basis for action:

Lack of good-quality water limits livestock production under grazing and rangeland conditions. Livestock production is also the cause of water contamination because of improper waste management in high-livestock-density zones. Leaching into aquifers should be reduced through better integration of livestock and crop production. Overgrazing has contributed to the degradation of grasslands and desertification. Where livestock is one among competing users of a limited water supply, the opportunity value of the water in question must be established to assist in determining the efficient and equitable allocation of water.

Strategy and programme targets:

Provision of appropriately spaced and reliable watering points and grazing lands within sustainable land-use systems. Animal water requirement is likely to increase from 60 billion litres per day to 65 billion litres per day by the year 2000.

Integration of land- and water-use strategies for the semi-arid areas to be developed in close cooperation with pastoral populations and based on clearly defined property rights.

Table 34. *Water for livestock*

Activities and related means of implementation	Level[1]	Considered by	
		ICWE[2]	UNCED[3]
1. Ensure the availability of good-quality water for livestock, taking into account their tolerance limits, with particular reference to salinity and toxic elements.	NPL	X	X
2. Increase the quantity of water sources available to livestock, in particular those in extensive grazing systems, in order both to reduce the distance travelled for water and to prevent overgrazing around water sources.	INPL	X	X
3. Prevent contamination of water sources with animal excrement in order to prevent the spread of diseases, in particular zoonoses.	INPL	X	X
4. Encourage water spreading schemes for increasing water retention of extensive grasslands to stimulate forage production and prevent runoff.	INPL	X	X
5. Encourage multiple use of water supplies through promotion of integrated agro–livestock-fishery–systems.	INPL	X	X

[1] Level of implementation: I=International; N=National; P=Provincial or sub-national; L=Local.
[2] Considered in ICWE *Report of the Conference* section 6.
[3] Considered in UNCED Agenda 21 paragraph: 18.65–81.

7.8 Inland fisheries and aquaculture

Basis for action:

Inland water fisheries should be managed to optimize the yield of aquatic food organisms through conservation of water quality and quantity and the functional morphology of inland aquatic systems. Aquaculture should be promoted as a component of integrated farming systems. At the same time, the development of aquaculture and coastal fisheries will need to be guided to protect the quality of aquatic systems and the environment.

Strategy and programme targets:

The major strategy consists of promoting inland fisheries and aquaculture within the framework of national and international water resources planning and management. In the context of competing uses for water the opportunity value should be applied as in the case of water for livestock. The targets are:

Table 35. *Inland fisheries and aquaculture*

Activities and related means of implementation	Level[1]	Considered by	
		ICWE[2]	UNCED[3]
1. Develop sustainable inland fishery management programmes compatible with multipurpose water resources planning and development.	INP	X	X
2. Study specific aspects of the hydrobiology and environmental requirements of key inland fish species in relation to varying water regimes.	I L	X	X
3. Prevent or mitigate modification of aquatic environments by other users or rehabilitate environments subjected to such modification on behalf of the sustainable use and conservation of biological diversity of living aquatic resources.	INPL	X	X
4. Establish and maintain adequate systems for the collection and interpretation of data on water quality and quantity and channel morphology related to the state and management of living aquatic resources, including fisheries.	IN	X	X
5. Develop and disseminate environmentally sound inland fisheries and aquaculture technologies that are compatible with local, regional and national water management plans and take into consideration social factors.	INPL	X	X
6. Introduce appropriate aquaculture techniques and related water development and management practices in countries not yet experienced in aquaculture.	IN	X	X
7. Assess environmental impacts of aquaculture with specific reference to commercialized culture units and potential water pollution from processing centres.	I PL	X	X
8. Evaluate economic feasibility of aquaculture in relation to alternative use of water, taking into consideration the use of water of marginal quality and investment and operational requirements.	INPL	X	X

[1] Level of implementation: I=International; N=National; P=Provincial or sub-national; L=Local.
[2] Considered in ICWE *Report of the Conference* section 6.
[3] Considered in UNCED Agenda 21 paragraph: 18.65–81.

- to increase capture fisheries from the current 7 million tonnes/year to 10 million tonnes/year by the year 2000.
- to double inland aquaculture production/year from the current 7 million to 14 million tonnes by the year 2000.

7.9 Capacity building

Basis for action:

Lack of efficient institutions and trained human resources has been a major cause of inefficient water management for agricultural and rural development. There is an urgent need for developing nations to build their own long-term capacities for integrated management of rural resources that support their communities. The actions at local, provincial, national and international levels will require an institutional framework and mechanisms for coordination within countries and between countries and the United Nations system of organizations and donor and financing agencies. These concepts have been elaborated in IHE-Delft/UNDP (1991).

The adoption of more efficient water use, protection of water quality from pollution by agricultural chemicals and other contaminating materials, and establishment of clearly defined property rights and obligations require the introduction of appropriate legal instruments at local and national levels. Given the need to address multi-sectoral problems related to water use at the rural level, inter-institutional linkages will need to be established. Strengthening the capacity of institutions to administer the legal, economic and monitoring functions is essential.

The importance of a functional and coherent institutional framework at the national level to promote water and sustainable agricultural development has generally been fully recognized at present. In addition, an adequate legal framework of rules and regulations should be in place to facilitate actions on agricultural water use, drainage, water-quality management, small-scale water programmes and the functioning of water-users' and fishermen's associations. Legislation specific to the needs of the agricultural water sector should be consistent with, and stem from, general legislation for the management of water resources.

Strategy and programme targets:

The major strategy consists of the creation of policy and legal frameworks, the development and strengthening of institutions, the dissemination of hydrological and other databases, the promotion of community participation and the training of human resources, all on a continuing basis.

A realistic national-level goal for the year 2000 may include establishing national policies and programmes to address institutional and human resource deficiencies.

Table 36. *Capacity building*

Activities and related means of implementation	Level[1]	Considered by	
		ICWE[2]	UNCED[3]
1. Improve water-use policies related to agriculture, fisheries and rural development and legal frameworks for implementing such policies.	NP	X	X
2. Strengthen institutions based upon human resources development and managerial systems through: (a) reviewing, strengthening and restructuring, if required, existing institutions in order to enhance their capacities in water-related activities, while recognizing the need to manage water resources at the lowest appropriate level; (b) reviewing and strengthening organizational structures, functional relationships and linkages between ministries and within individual ministries; (c) providing specific measures that require support for institutional strengthening, *inter alia*, through long-term programme budgeting, staff training, incentives, mobility, equipment and coordination mechanisms.	NPL	X	X
3. Assess training needs for agricultural water management and increase formal and informal training activities through: (a) developing practical training courses for improving the ability of extension services to disseminate technologies and strengthen farmers' capabilities, with special reference to small-scale producers; (b) training staff at all levels, including farmers, fishermen and members of local communities, with particular emphasis on the training of women; (c) increasing the opportunities for career development to enhance the capabilities of administrators and officers at all levels involved in land- and water-management programmes.	NP	X	X
4. Enhance the involvement of the private sector, where appropriate, in human resource development and provision of infrastructure.	NP	X	X
5. Transfer existing and new water-use technologies by creating mechanisms for cooperation and information exchange among national and regional institutions.	INP	X	X

[1] Level of implementation: I=International; N=National; P=Provincial or sub-national; L=Local.
[2] Considered in ICWE *Report of the Conference* section 6.
[3] Considered in UNCED Agenda 21 paragraph: 18.65–81.

Education and training of human resources should be actively pursued at the national level through: (a) assessment of current and long-term human resources management and training needs; (b) establishment of a national policy for human resources development; and (c) initiation and implementation of training programmes for staff at all levels as well as for farmers.

7.10 Targets and costs

(i) Targets

(a) To achieve 15.2 million hectares of new irrigation development in 130 developing countries by the year 2000;
(b) To improve and modernize 12 million hectares of irrigated land in 130 developing countries by the year 2000;
(c) To establish small-scale water programmes for the improvement of a total of 10 million hectares of rain-fed arable land in 130 developing countries by the year 2000;
(d) To provide improved drainage and reduction of excessive losses to groundwater on 7 million hectares of irrigated land in 130 developing countries by the year 2000;
(e) To increase capture fishery production from 7 million to 10 million tonnes/year, and inland aquaculture production from 7 million to 14 million tonnes/year by the year 2000.

(ii) Cost estimates

The estimated average total annual cost (for 1993–2000) of implementing the activities of this programme is about US$ 13.2 billion, including about US$ 4.5 billion from the international community on grant or concessional terms.

A breeding-ground for diarrhoeal diseases: a Bombay slum where waste-water is in contact with water for drinking and washing. Credit: WHO/S.K. Dutt.

8

Drinking water supply and sanitation (WSS)

8.1 Background

The provision of wholesome supplies of drinking water, together with effective sanitation systems, which began in the mid-nineteenth century, is one of the key components of modern living. It can be argued that the cities and towns which possess such services, together with their surrounding rural areas, approach most closely to the state of sustainable development. Settlements without these services could be considered the farthest from it in terms of water resources.

Safe water supplies and environmental sanitation are vital for protecting the environment, improving health and alleviating poverty (cf. Horchani, 1992). Safe water is also crucial to many traditional and cultural activities. An estimated 80% of all diseases and over one third of deaths in developing countries are related to the consumption of contaminated water, and on average as much as one tenth of each person's productive time is sacrificed to water-related diseases. Concerted efforts during the 1980s brought water and sanitation services to hundreds of millions of the world's poorest people. The most outstanding of these efforts was the launch in 1981 of the International Drinking Water Supply and Sanitation Decade (IDWSSD), which resulted from the Mar del Plata Action Plan adopted by the United Nations Water Conference in 1977. The commonly agreed premise was that 'all peoples, whatever their stage of development and their social and economic conditions, have the right to have access to drinking water in quantities and of a quality equal to their basic needs' (UN, 1977). The target of the Decade was to provide safe drinking water and sanitation to underserved urban and rural areas by 1990, but even the unprecedented progress achieved during the Decade was not enough (cf. WHO, 1991). One in every three people in the developing world still lacks these two most basic requirements for health and dignity. It is also recognized that human excreta and sewage are important causes of the deterioration of water quality in developing countries, and the introduction of available technologies, including appropriate technologies, and the construction of sewage treatment facilities could bring significant improvement.

The New Delhi Statement (adopted at the Global Consultation on Safe Water and Sanitation for the 1990s, which was held in New Delhi from 10 to 14 September 1990) formalized the need to provide, on a sustainable basis, access to safe water in sufficient quantities and proper sanitation for all, emphasizing the 'some for all rather than more for some' approach (cf. UNDP, 1991). Four guiding principles were agreed:

(a) Protection of the environment and safeguarding of health through the integrated management of water resources and liquid and solid wastes;
(b) Institutional reforms promoting an integrated approach and including changes in procedures, attitudes and behaviour, and the full participation of women at all levels in sector institutions;
(c) Community management of services, backed by measures to strengthen local institutions in implementing and sustaining water and sanitation programmes;
(d) Sound financial practices, achieved through better management of existing assets, and widespread use of appropriate technologies.

Past experience has shown that specific targets should be set by each individual country. At the World Summit for Children, in September 1990, Heads of State or Government called for both universal access to water supply and sanitation and the eradication of guinea worm disease by 1995. Even for the more realistic target of achieving full coverage in water supply by 2025, it is estimated that annual investments must reach double the current levels. One realistic strategy to meet present and future needs, therefore, is to develop lower-cost but adequate services that can be implemented and sustained at the community level.

To ensure the feasibility, acceptability and sustainability of planned water supply services, adopted technologies should be responsive to the needs and constraints imposed by the conditions of the community concerned. Thus, design criteria should involve technical, health, social, economic, institutional and environmental factors that determine the characteristics, magnitude and cost of the planned system. Relevant international support programmes should address the developing countries concerning:

(a) Pursuit of low-cost scientific and technological means, as far as practicable;
(b) Utilization of traditional and indigenous practices, as far as practicable, to maximize and sustain local involvement;
(c) Assistance to country-level technical/scientific institutes to facilitate curricula development to support fields critical to the water and sanitation sector.

International standards for drinking water are under consideration by the World Health Organization. Drinking water standards have been set for many countries in Europe and North America. In these countries biologically safe water has, for the most part, been attained and attention is now focused more on organic and inorganic contaminants which are proving to be a great challenge for water treatment technologies. The standards set in North America and Europe are illustrated in Sayre (1988). They are being considered for application in other countries, but the biological standards, especially, may have to be less stringent in most of the developing countries.

8.2 Providing water supply and sanitation coverage

Basis for action:

Despite many commendable efforts, the International Drinking Water Supply and Sanitation Decade left 244 million urban residents without access to safe water supplies and 380 million lacking adequate sanitation. While an additional 1141 million rural residents obtained access to safe water supplies, an estimated 842 million remained unserved in 1990. Only 427 million rural residents obtained access to appropriate means of excreta disposal, leaving a total of around 1385 million still unserved. These unserved millions are poor, politically powerless, and living in dismal conditions. The situation is aggravated still further by large numbers of broken down and malfunctioning water and sanitation systems.

The global population is expected to reach 8 billion by the year 2020. The consequent dramatic growth in food demand and the pressures upon the natural resource base pose an enormous challenge to the rural sector, as it will not only have to increase food production substantially, but, while being the major user of water resources, it will have to release freshwater resources to meet increasing demands for domestic, industrial and ecosystem management purposes. Additionally, the rural sector will have to contribute significantly to the conservation of the natural resource base.

Accelerated provision of basic water and sanitation services is a prerequisite for improved health and for sustainable social and economic advancement. The poor put a high priority on the dignity and convenience of clean water and hygienic sanitation, reflected in a proven willingness to pay a substantial proportion of their income for reliable services, (cf. Bhatia and Falkenmark, 1992). To them and to many of the world's more fortunate inhabitants poverty is highly correlated with lack of water services (cf. Traoré, 1992).

Problems encountered in implementing programmes to achieve the IDWSSD goals include the lack of sufficient funding, inadequate cost-recovery to ensure sector sustainability, insufficient trained manpower, poor operation and maintenance of systems, and lack of community involvement. In addition, improper utilization of systems resulting from lack of awareness of the health consequences of

Table 37. *Providing water supply and sanitation coverage*

Activities and related means of implementation	Level[1]	Considered by	
		ICWE[2]	UNCED[3]
1. Provide international technical and financial support to develop costed proposals (including operation and maintenance) for providing services to the urban and rural poor, and allocate commensurate budgets.	IN	X	
2. Adopt appropriate procedures for project implementation and the improvement of WSS services through: (a) designing of projects on the basis of effective demand and what people want and are willing to pay for; (b) utilizing technologies appropriate to the capabilities of the communities, for example the village level operation and maintenance (VLOM) programme of the World Bank; (c) adopting cost-recovery policies which are compatible with the users' willingness to pay for WSS services; (d) adjusting land-use practices and improving sanitation and community and industrial waste disposal to ensure the protection of groundwater and surface water including coastal seas.	INPL	X	X
3. Promote community ownership, empowerment and management of WSS systems through: (a) accepting of WSS as a comprehensive development activity rather than an engineering project; (b) actively promoting the use of participatory mechanisms for involving communities, with emphasis on the role of women, in the planning and implementation of programmes for water use, water conservation, and WSS services.	INPL	X	X
4. Ensure the equitable and efficient use of water through: (a) using progressive water tariffs that reflect the true cost of providing and maintaining supplies, including cost-recovery, and encouraging conservation and minimization of waste; (b) charging policies that will enable the very poor to receive basic services; (c) basing the choice of technology and service levels on user preferences and willingness to pay; (d) introducing suitable cost-recovery through demand management mechanisms.	NPL	X	X
5. Prepare WSS sector plans and national programmes with particular emphasis on the optimization and sustainability of existing systems, the adoption of appropriate technology in new schemes, sound financial practices, and the extension of coverage to marginal areas.	NP	X	X

[1] Level of implementation: I=International; N=National; P=Provincial or sub-national; L=Local.
[2] Considered in ICWE *Report of the Conference* paragraphs 5.17–22; 6.14–20.
[3] Considered in UNCED Agenda 21 paragraph: 18.47–55.

unhygienic services has often led to a less than satisfactory level of health benefits achieved after the commissioning of a system.

Strategy and programme targets:

Provision should be made for reliable water, sanitation, solid waste and drainage services to the poor as a priority component of national environmental management strategies, involving the private sector and non-governmental organizations. This will imply a strengthening of the water supply and sanitation sector with emphasis on institutional development, efficient management and an appropriate framework for financing the services.

In urban areas, WSS services should be extended, with the aim of reducing the number lacking services at the end of the International Drinking Water Supply and Sanitation Decade in 1990, to half that number by the year 2000 and to provide coverage for all by 2015.

In rural areas the sustainability of WSS services should be ensured through sector strengthening with emphasis on the implementation of institutional development programmes based on efficient management and an appropriate framework for financing of the services. This will require the awarding of high priority to sector plans for drinking water and sanitation.

The target should be that, by the year 2000, all countries and rural communities should:

(a) Have agricultural and water resource systems so that, through local production and commodity trading, they will have access to sufficient food to meet their basic nutritional needs;
(b) Have access to safe water in sufficient quantities and adequate sanitation to meet their health needs and maintain the essential qualities of their local environments;
(c) Develop their water resource systems only as components of comprehensive integrated programmes designed to provide for long-term sustainment of human welfare and the careful management of natural ecosystems.

8.3 Health impacts

Basis for action:

The risk factors involved in the lack of adequate water supply and sanitation cover a broad range, contact with contaminated water (for example schistosomiasis); protozoal, bacterial and viral contamination of water (for example amoebic dysentery, various enteric infections, hepatitis), and the propagation of guinea worm infection. Unreliable water supplies force people to store drinking water in and around houses, which may create breeding sites for Aedes vectors of dengue

Table 38. *Health impacts*

Activities and related means of implementation	Level[1]	Considered by	
		ICWE[2]	UNCED[3]
1. Provide technical and financial support for the design and implementation of programmes in developing countries through: (a) intensification of water quality control and the operation of water treatment plants and other supplies (wells, springs, etc.); (b) assistance to communities to upgrade water storage facilities and identify appropriate technologies for the disinfection of water and wastes, promoting the use of in-house disinfecting techniques where public supplies cannot be adequately safeguarded.	IN	X	X
2. Promote environmental health through: (a) establishment of protected areas for sources of drinking-water supply; (b) sanitary disposal of excreta and sewage, using appropriate systems to treat waste waters in urban and rural areas; (c) expansion of urban and rural water-supply and development and expansion of rainwater catchment systems, particularly on small islands, in addition to the reticulated water-supply system; (d) treatment and safe reuse of domestic and industrial waste waters in urban and rural areas; (e) development of low-cost sewerage and disinfection systems for low-income settlements and make low-cost water supply and sanitation technology choices available.	NPL	X	X
3. Ensure that the monitoring of relevant health indicators accompanies any water supply and sanitation project.	NPL	X	X
4. Introduce water supply and sanitation components into all irrigation projects and adopt wide-scale environmental management measures to control disease vectors.	INPL	X	X
5. Promote specific solutions as appropriate, for example: (a) the use of larvivorous fish for malaria vector control in village ponds; (b) in areas with high levels of diarrhoea and other enteric infection, focus the WSS activities on sanitation improvement and water quality monitoring.	INPL	X	X
6. Develop and implement multisectoral rapid response interventions for dealing with cholera outbreaks.	IN	X	X

[1] Level of implementation: I=International; N=National; P=Provincial or sub-national; L=Local.
[2] Considered in ICWE *Report of the Conference* paragraphs 5.17–22; 6.14–20.
[3] Considered in UNCED Agenda 21 paragraph: 18.47–55.

haemorrhagic fever. Ponds and collections of seepage water around hand pumps may contribute to malaria problems. Inadequate water and sanitation is responsible for 90% of cholera outbreaks. Poor populations on the fringe of large cities are particularly exposed to contamination in such outbreaks. Rigorous care in drinking and eating habits and in personal hygiene, is the most effective way of reducing cholera risk.

Strategy and programme targets:

The strategy consists of hygiene education and elimination of transmission foci; adoption of appropriate technologies for water treatment; and wide-scale adoption of environmental management measures to control disease vectors.

Medium- and long-term plans for environmental sanitation should be established by 1995 to ensure permanent protection of vulnerable groups against disease risks, especially cholera.

Consideration should be given to the creation of an international fund to respond to health-hazard emergencies.

8.4 Capacity building

Overall national capacity building at all administrative levels, involving institutional development, coordination, human resources, community participation, health and hygiene education and literacy, has to be developed according to its fundamental connection both with any efforts to improve health and socio-economic development through water supply and sanitation and with their impact on the human environment. Capacity building should therefore be one of the underlying keys in implementation strategies, as elaborated in IHE-Delft/UNDP (1991). Institutional capacity building should be considered to have an importance equal to that of the sector supplies and equipment component so that funds can be directed to both. This can be undertaken at the planning or programme/project formulation stage, accompanied by a clear definition of objectives and targets. In this regard, technical cooperation among developing countries building on their available wealth of information and experience and the need to avoid 'reinventing the wheel', is crucial. Such a course has already proved cost-effective in many country projects.

To plan and manage water supply and sanitation effectively at the national, provincial, district and community level, and to utilize funds most effectively, trained professional and technical staff must be developed within each country in sufficient numbers. To do this, countries must establish manpower development plans, taking into consideration present requirements and planned developments.

Table 39. *Capacity building*

Activities and related means of implementation	Level[1]	Considered by	
		ICWE[2]	UNCED[3]
1. Develop policies, strategies and legislation to ensure an equitable allocation of resources through: (a) establishment of explicit national policies and development plans for the WSS sector; (b) formulation of legislation necessary to support viable water and sanitation systems; (c) consideration of health consequences in all water resource programmes in order to enhance social and economic development.	INPL	X	X
2. Promote institutional development through: (a) strengthening the functioning of governments in water resources management and, at the same time, giving full recognition to the role of local authorities; (b) application of the principle that decisions are to be taken at the lowest appropriate level, with public consultation and involvement of users in the planning and implementation of water projects; (c) integration of community management of water within the context of overall planning; (d) development of linkages between national water plans and community management of local waters; (e) support and assistance to communities in managing their own systems on a sustainable basis; (f) encouragement of water development and management based on a participatory approach, involving planners and policy makers at all levels and the local population, especially women, youth, indigenous people; (g) encouragement of local water associations and water committees to manage community water supply and sanitation systems and provision of technical backup; (h) strengthening of local health services; (i) strengthening the monitoring and reconnaissance capabilities of water authorities in relation to health risks; (j) development and implementation of multisectoral rapid response interventions for dealing with epidemics, especially cholera outbreaks.	INPL	X	X
3. Promote human resource development at all levels, including special programmes for women through: (a) promotion of primary health and environmental care at the local level, including training for local communities in appropriate water management techniques and primary health care; (b) assistance to service agencies in becoming more cost-effective and responsive to consumer needs; (c) mobilization and facilitation of the active involvement of women in water management teams.	INPL	X	X
4. Promote health education programmes directed towards: (a) enhancement of hygiene, with a focus on women and children, to stimulate demand for and use of improved sanitation and	NPL	X	X

waste disposal facilities; (b) education of populations at risk, particularly regarding their individual responsibilities in the prevention and transmission of cholera and other water-related diseases and regarding the opportunities of using suitable low-cost water supply and sanitation technology; (c) promotion of safe disposal/use of human excreta/waste; (d) activating mechanisms to provide populations with better access to essential technical and environmental information, in order to strengthen their role in decision-making processes; (e) urging the general population as well as public and private institutions to intensify sanitary controls and to improve the cleaning and disinfecting of wells, tanks and drinking water installations.			
5. Promote better and more efficient techniques through: (a) rehabilitation of defective systems, reduction of wastage and safe reuse of water and wastewater; (b) programmes for rational and efficient water use and ensured operation and maintenance; (c) research and development of appropriate technical solutions; (d) substantially increasing urban wastewater treatment capacity commensurate with increasing loads.	NPL	X	X
6. Strengthen monitoring, evaluation and information management at national and sub-national levels through: (a) evaluation of water and sanitation facilities, including institutional arrangements for operation, maintenance, planning and financing of these services; (b) identification of a core set of WSS indicators and establishment of national monitoring offices; (c) implementation of annual monitoring of WSS indicators; (d) use of indicators at regional and global levels to promote awareness and raise funds; (e) processing, analyzing and publication of monitoring results at national and local levels as a management and advocacy/awareness creation tool; (f) improvement of coordination, planning and implementation, with the assistance of improved monitoring and information management, particularly in community-based self-help projects; (g) development of national inventories of communities at risk, collection of environmental health and epidemiological data, selection of priority areas, and implementation of appropriate interventions.	NP	X	X
7. Develop international support mechanisms to encourage international cooperation and to encourage national governments and external support agencies to develop common policies through: (a) promotion of the exchange of information between water agencies from different countries; (b) coordination and expansion of international efforts regarding technical and financial cooperation to developing countries, particularly related to sector sustainability; (c) development of guidelines for the sustainability of WSS services to be used by external support agencies and developing countries in the formulation of development plans; (d) channelling WSS assistance in conformity with national policies.	IN	X	X

[1] Level of implementation: I=International; N=National; P=Provincial or sub-national; L=Local.
[2] Considered in ICWE *Report of the Conference* paragraphs 5.17–22; 6.14–20.
[3] Considered in UNCED Agenda 21 paragraph: 18.47–55.

Subsequently, the development and performance of country-level training institutions should be enhanced so that they can play a pivotal role in capacity building. It is also important that countries provide adequate training for women in the sustainable maintenance of equipment, water resources management and environmental sanitation.

The implementation of water supply and sanitation programmes is a national responsibility. To varying degrees, responsibility for the implementation of projects and the operating of systems should be delegated to all administrative levels down to the community and individual served. This also means that national authorities, together with the agencies and bodies of the United Nations system and other external support agencies providing support to national programmes, should develop mechanisms and procedures to collaborate at all levels. This is particularly important if full advantage is to be taken of community-based approaches and self-reliance as tools for sustainability. This will entail a high degree of community participation, involving women, in the conception, planning, decision making, implementation and evaluation connected with projects for domestic water supply and sanitation.

Strategy and programme targets:

Strengthening national capacities to plan, implement, and monitor integrated water management programmes. The major strategy is to create policy and legal frameworks on a participatory basis, as well as develop and strengthen institutions at all levels. This should be accomplished with emphasis on community participation and human resource development, taking into consideration the full involvement of women, who constitute a substantial proportion of the world's farmers.

Resolving competing demands on water resources through the application of appropriate economic, legal and institutional mechanisms. This objective is achievable through better integration of water use within the framework of overall national economic, agricultural and environmental policies and adoption of demand management strategies, ranging from appropriate pricing policies to implementation of relevant legal frameworks and entitlements supported by the widespread introduction of water saving technologies.

8.5 Targets and costs

(i) Targets

(a) To expand the coverage of community water supply and sanitation services to an additional 1.6 billion people by the year 2000;

(b) To achieve full coverage in water supply by the year 2025.

(ii) Cost estimates

Accelerated development is necessary to reach the desired coverage of water supply and basic sanitation services by the year 2000. The rate of investment for the years until 2000 would have to be doubled to a total of US$ 20 billion annually if complete service coverage were to be reached. The external component should be maintained at no less than one third of this, i.e. at about US$ 6.7 billion annually. The improved operation, maintenance and management of systems and the full utilization of the investments made requires the allocation of external support of the order of US$ 0.7 billion. The total external funding needs, on grant or concessional terms, until the year 2000 are, therefore, US$ 7.4 billion annually*.

* The New Delhi Statement concludes that "if costs were halved and financial resources at least doubled, universal coverage could be within range by the end of the century".

Like all development problems, the issue of water supply and sanitation is multidimensional and interconnected, with a problem existing in one area, influencing the evolution and outcome in others. Hence the need for building the capacity to manage the resource better. Credit: 'Women, water supply and sanitation', ILO, UN-INSTRAW, UN-DCTD.

9
Capacity building

9.1 Background

The purpose of this chapter is to bring together and discuss the several components included under the overall term 'capacity building'. In doing this heavy reliance has been placed on the deliberations of participants at the UNDP Symposium held at Delft, The Netherlands, from 3–5 June 1991. This Symposium resulted in the Delft Declaration and a book, *A Strategy for Water Sector Capacity Building* (IHE-Delft/UNDP, 1991), the recommendations from which formed the basis for much deliberation at ICWE and for many of the recommendations at UNCED.

There is now a strong consensus that institutional weaknesses and malfunctions are major causes of many national water services being ineffective. There is an urgent requirement in many countries for attention to be given to building and developing institutional capacities. Capacities have to be built in order to address the needs relevant to all water sector activities, including water resources assessment and water planning and management. Capacities must be built to allow individual countries to look after their own affairs. Capacities at all levels of government, from national through provincial or regional to local levels of government, must be developed in order that total needs be met.

Capacity building is a cross-sectoral activity, being needed within both the urban and the rural areas, being needed to address environmental and health concerns and being needed in order to implement effectively policies for sustainable food production, and so on. The need for capacity building is addressed in all parts of the chapter on freshwater resources of the UNCED Agenda 21 document (UN, 1993). Capacity building is referred to throughout the Report of the Dublin Conference (WMO, 1992) and it has been an integral part of all the chapters in this book.

Countries and External Support Agencies (ESAs) increasingly recognize the importance of the capacity building process for sustainable development at national, provincial and local levels. While opinion is still divided in defining the scope of the term 'capacity building', there is agreement that it includes three basic elements:

(a) At the **sectoral level** the creation of an enabling environment with appropriate policy and legal frameworks;
(b) At the **institutional level** the development of planning and management processes to allow collective skills to be used effectively;
(c) At the **individual level** the comprehensive development of human resources and the promotion of awareness of water issues.

In implementing capacity building processes there is recognition that circumstances vary between countries. Thus, while all elements of the capacity building process must be addressed in all countries, each country will need a programme tailor-made to its own requirements. While geographic regions often contain countries with a certain commonality of circumstances and conditions, there may still be significant political and socio-economic differences between those countries, thus necessitating significantly different programmes of capacity building.

There is also a general recognition and acceptance that each country must be responsible for its own affairs. Requests for assistance from outside must be initiated within the country. External assistance, in whatever form, must promote the development of internal self-sufficiency. The timetable to achieve self-sufficiency in water management may be spread over years or decades, but striving for its eventual attainment is of fundamental importance.

9.2 The creation of an enabling environment at the sectoral level

Basis for action:

In creating the enabling environment for lowest-appropriate-level management, the role of Government includes mobilization of financial and human resources, legislation, standard-setting and other regulatory functions, monitoring and assessment of the use of water and land resources, and creating of opportunities for public participation.

While it is recognized that each country will have its own political and socio-economic history, which may constrain the flexibility of approach, institutional structures at the national level are nearly always of fundamental importance in determining the efficiency of the whole management system within a country. Linkage from the national through the provincial to the local levels of government is also, however, particularly important within the water sector as so many important practical day-to-day decisions regarding the resource must be made at very local levels.

Water resources planning and management should be an integral part of overall national economic planning. The objectives and strategies for the water sector should be derived from the national planning process. Thus, it is of prime importance that all major water issues be incorporated within national plans. Without such inclusion, water issues may be marginalized and their true importance may not be realized.

It is also of great importance that there be proper linkage between the various elements of the water sector. It is common that water issues are divided between the jurisdictions of various government departments – energy, agriculture, industry, and so on – and that communication between departments is inadequate. This subverts the proper communication necessary for effective integrated management of the resource.

The initiation, promotion and development of integrated water planning and management practices can only take place effectively and efficiently if overall institutional structures and appropriate legislative frameworks are in place. Thus, appropriate legislation must be enacted to define responsibilities and obligations of the various sectors of government and to define the basis of devolvement of responsibility from the central to the local levels of government. In cases of transboundary water courses, administrative structures and legislation may be particularly difficult to agree upon.

Existing administrative structures will often be quite capable of achieving effective local water resources management, but the need may arise for new institutions based upon the perspective, for example, of river catchment areas, district development councils and local community committees. Although water is managed at various levels in the socio-political system, demand-driven management requires the development of water-related institutions at appropriate levels, taking into account the need for integration with land-use management.

In those cases where structural changes are required, there must always be the recognition that existing structures and existing legislation can seldom be replaced within short time frames. Rather, the process of change must normally be slow and must build on existing structures. The acceptance of proposals for changing or modifying existing institutional structures or legislative frameworks must always be the prerogative of the national government concerned.

Capacity building is, indeed, a long-term, continuous process which should be phased to accommodate requirements of national governments and ESAs. Each individual phase should have clearly defined and measurable targets. A necessary prerequisite, therefore, is the setting of realistic and achievable goals based on the available resources.

International agencies and donors have an important role to play in providing support to developing countries in creating the required enabling environment for integrated water resources management. This should include, as appropriate,

Table 40. *The creation of an enabling environment at the sectoral level*

Activities and related means of implementation	Level[1]	Considered by	
		ICWE[2]	UNCED[3]
1. Initiation, promotion and maintenance of assessment processes to assess (a): the capabilities and needs of agencies at all levels to design and implement sustainable projects; (b) the technical, institutional and budgetary requirements for planning, design, implementation, operation and maintenance of new projects; (c) the extent of coordination between water related agencies and the need for better cooperation; (d) the status of the legislative and policy frameworks at all levels; (e) the efficiency of programmes through, *inter alia*, comparison of costs and options for achieving identified targets; (f) information needs.	N	X	X
2. Promotion of enabling legislation to: (a) incorporate sustainable programmes for the water sector in national development plans; (b) enable integrated planning and management; (c) enable management decisions to be made at the lowest appropriate level; (d) make practical provision for the long-term sustainability of WRA and forecasting activities; (e) adopt regulatory and economic instruments to combat pollution from all sources, and ensure that the monitoring and surveillance capability is available to enforce standards; (f) implement national programmes to introduce sanitary waste disposal facilities based on low-cost upgradeable technologies.	N	X	X
3. Promotion of cooperation through: (a) the establishment of coordinating mechanisms; (b) the transfer of technology; (c) the rationalization of public and private sector intervention; (d) the establishment of inter-sectoral planning groups involving housing, land management and environmental protection agencies at city and regional level, and health, finance and environment ministries at national level.	IN	X	X
4. Acquisition of funding in order to ensure that all countries have at least basic capabilities and services in the water sector.	IN	X	X

[1] Level of implementation: I=International; N=National; P=Provincial or sub-national; L=Local.
[2] Considered in ICWE *Report of the Conference* sections 2–6.
[3] Considered in UNCED Agenda 21 paragraph: 18.13–22; 18.28–34; 18.41–46; 18.51–55; 18.60–64; 18.77–81; 18.86–90.

donor support to local levels in developing countries, including community-based institutions, non-governmental organizations and women's groups.

Strategy and programme targets:

The overall goal should be to have in place institutional and legislative structures at the national level which would allow completely integrated planning and management to take place in a sustainable manner and with great efficiency.

At the sectoral level three important strategies may be identified:

Strategy 1. The initiation and maintenance of assessment processes.

An essential prerequisite to the promotion of better institutional and legislative arrangements as well as the development of planning and management policies is a series of appropriate assessments. Such assessments would include not only a review of the characteristics and state of the resource itself, but also a review of the enabling environment, including cooperative arrangements, institutional status, and so on. The process of assessment should be continuous as circumstances are in a state of constant flux.

Strategy 2. The promotion of enabling legislation.

Appropriate legislation must be in place to define jurisdictions, responsibilities and obligations within management structures. This is essential to allow necessary cooperation to take place. This in turn would allow management policies to be initiated which would recognize:

(a) The need for development to be sustainable;
(b) The need to harmonize environment and development issues;
(c) The need for planning and management to be integrated internally and with all sectors of socio-economic activity.

Strategy 3. The promotion of cooperation.

Cooperation is essential on a number of levels.

(a) Within countries there must be cooperation between agencies within each level of government and between different levels of government. There must also be cooperation between governments and the private sector and between governments and non-governmental organizations.
(b) At the international level there are needs for cooperation of various types:
Cooperation between countries sharing transboundary waters through international river basin authorities.
Bilateral and multilateral cooperation between countries for the sharing of costs and the sharing of technical and managerial knowledge.

Cooperation between international organizations, including UN agencies and NGOs and individual countries.

The modalities for such varied forms of cooperation are numerous and there is always a need that all forms of cooperation are appropriately coordinated to avoid inefficient overlap of activities. Governments should be urged to coordinate the activities of ESAs and ESAs themselves should be encouraged to integrate and coordinate their efforts as much as possible.

ESAs should be encouraged to adopt capacity building as an essential element of their assistance efforts. They should be encouraged, when appropriate, to establish linkages not only with national governments, but also with lower levels of government.

9.3 The development of planning and management processes at the institutional level

Basis for action:

The institutional framework required for the planning and management of water resources is nation specific and should develop from the existing situation. Due to the inherent complexities in water resources management, it is rarely feasible or wise to allocate all necessary research and management functions to one institution. Planning, decision making, technical implementation, etc. have to be distributed among institutions at various levels. The coordination of functions from grass roots to central level is therefore an important element in achieving integrated water resources management, and may need strengthening.

Many national and local institutions responsible for water management and water service delivery do not work efficiently or effectively because of a number of factors including:

- inappropriate policies for water management;
- unclear definition of the mandates of the institutions;
- a working environment that is not conducive to job satisfaction resulting in inefficient institutional functioning;
- lack of participation and linkage with communities and customers resulting in a lack of commitment;
- lack of resources (inadequate funding and human resources);
- inadequate education and training facilities.

Capacity building activities should be undertaken both within and among water subsectors to provide improved coordination. The collaboration between subsectors is becoming increasingly important and urgent in dealing with water availability and quality issues created by competing users, in particular with regard to demand from large urban areas and irrigation, and pollution caused by users. To this end, there is a need to establish control and coordination between institutions at national as well as local levels.

Administrative rules and regulations governing the functioning of institutions are often unsuitable and may need overhauling to enable the institutions to act in a more flexible manner. In particular, the mandates of the institutions within the overall national framework often are in need of rationalization. Rules and regulations in support of integrated water resources management often need to be developed and expanded and require a continuous process of review.

In many countries a continued need exists for governments to adopt an appropriate community participation approach, including the creation of community organizations and the building of adequate staff capacity to this end.

Strategy and programme targets:

An important basic principle which should govern the development of strategies and programme targets is that every country should aim to be self-sufficient in addressing its own institutional needs. All international aid should be directed towards this goal. In addition, wherever possible, traditional and indigenous practices and expertise should be used and incorporated into institutional functioning.

Strategy number 1

An assessment of institutional needs should be undertaken at the national, provincial and local levels (and in the cases of transboundary basins, at the international level) to identify weaknesses within the institutions themselves and weaknesses in the linkages between institutions.

Strategy number 2

Development of the capacities of individual institutions.

Strategy number 3

The promotion of cooperation and coordination between institutions at all levels of operation. Linkages should also be promoted between institutions and the private sector, communities and customers.

Strategy number 4

The promotion and development of international linkages. This should be undertaken between governments, with international organizations including NGOs and between institutions themselves including universities. These linkages would facilitate technology transfer and information exchange and would promote large-scale joint and multidisciplinary programmes.

Table 41. *The development of planning and management processes at the institutional level*

Activities and related means of implementation	Level[1]	Considered by	
		ICWE[2]	UNCED[3]
1. Assessment of institutional needs and capabilities at the national, provincial and local levels in order to establish an institutional framework which is efficient and functional and which ensures that the real needs are addressed. Identification of critical areas for water-related adaptive research.	NPL	X	X
2. Development of the capacities of individual institutions through: (a) upgrading and maintenance of laboratory services; (b) upgrading facilities and procedures for storing and safeguarding hydrological data; (c) promoting the development of new technologies; (d) strengthening the monitoring and reconnaissance capabilities of water authorities; (e) building up the financial viability of municipal water and sewerage agencies; (f) strengthening of extension services and adapting research results to management requirements; (g) ensuring that a minimum percentage of funds for water resource development projects is allocated to research and development, particularly in externally funded projects.	NPL	X	X
3. Promotion of linkages between institutions within countries through: (a) improving networks of stations to meet accepted standards and guidelines for the provision of hydrological data; (b) establishing intersectoral programmes and coordination mechanisms which integrate land-use planning with water-use and conservation requirements; (c) promoting inter-laboratory comparison studies and establishment of regional reference laboratories or national/regional equipment service centres.	NPL	X	X
4. Promotion of linkages between institutions and the private sector, communities and customers through: (a) encouraging and equipping local water associations and water committees to manage community water supply systems; (b) enabling private sector agencies, including NGOs, to offer support services, where these can be provided more efficiently than through public utilities; (c) combining the skills of NGOs, the private sector, and the users themselves in the planning, implementation and maintenance of water and	NPL	X	X

sanitation systems; (d) encouraging local dialogues involving WSS agencies, community groups, NGOs and private sector organizations; (e) ensuring that water, sanitation and waste disposal programmes respond to consumer demands through community involvement in decision making; (f) enhancing community participation and ensuring the role of women in community-based water projects; (g) enhancing involvement of the private sector in human resources development and infrastructure; (h) utilization of traditional and indigenous practices, as far as practicable, to maximize and sustain local involvement.			
5. Promotion of international linkages through: (a) establishing and strengthening research and development programmes appropriate to the needs of regions or countries; (b) building the capacity for hydrological research in developing countries with regard to integrated water resources management; (c) promoting the transfer of technologies and technical cooperation among local groups and national and international institutions; (d) encouraging information exchange; (e) encouraging involvement in cooperative programmes such as GEMS/WATER; (f) encouraging the establishment of twinning arrangements between institutions; (g) encouraging participation in joint/multidisciplinary programmes.	IN	X	X

[1] Level of implementation: I=International; N=National; P=Provincial or sub-national; L=Local.
[2] Considered in ICWE *Report of the Conference* sections 2–6.
[3] Considered in UNCED Agenda 21 paragraph: 18.13–22; 18.28–34; 18.41–46; 18.51–55; 18.60–64; 18.77–81; 18.86–90.

Table 42. *The development of human resources at the individual level and the promotion of awareness of water issues*

Activities and related means of implementation	Level[1]	Considered by	
		ICWE[2]	UNCED[3]
1. Assessment of current and long-term human resources management and training needs specific to the particular needs of individual countries.	NPL	X	X
2. Establishment of national policies for human resources development.	N	X	X
3. Initiation and implementation of training programmes for staff in all sectors and at all levels particularly programmes involving skills in community participation, low-cost technologies, and financial management..	INPL	X	X
4. Establishment and strengthening of education and training programmes on water-related topics, within an environmental and developmental context, for all categories of staff involving both men and women.	INPL	X	X
5. Development of sound recruitment, personnel and pay policies for staff of national and local water agencies.	NPL	X	X
6. Development of practical training courses for improving the ability of extension services to disseminate technologies and strengthen capabilities.	INPL	X	X
7. Increasing the opportunities for career development to enhance the capabilities of administrators and officers at all levels involved in land- and water-management programmes.	NPL	X	X
8. Undertaking initiatives directed towards politicians, decision-makers and the general public to raise awareness of the finite and fragile character of the water resources. The public-awareness campaigns will also include components to overcome user resistance by emphasizing the benefits of reliability and sustainability.	INPL	X	X

[1] Level of implementation: I=International; N=National; P=Provincial or sub-national; L=Local.
[2] Considered in ICWE *Report of the Conference* sections 2–6.
[3] Considered in UNCED Agenda 21 paragraph: 18.13–22; 18.28–34; 18.41–46; 18.51–55; 18.60–64; 18.77–81; 18.86–90.

9.4 The development of human resources at the individual level and the promotion of awareness of water issues

Basis for action:

Integrated water resources management requires the establishment and maintenance of a body of well-trained and motivated staff. Education and training programmes designed to ensure an adequate supply of trained personnel should be established or strengthened at the local, national, sub-regional or regional level. In addition, the provision of attractive terms of employment and career paths for professional and technical staff should be encouraged. Human resource needs should be monitored periodically, including all levels of employment. Plans have to be established to meet these needs.

The developmental work and innovation depend for their success on good academic training and staff motivation. International projects can help by enumerating alternatives, but each country needs to establish and implement the necessary policies and to develop its own expertise in the scientific and engineering challenges to be faced. The latter requires a body of dedicated individuals who are able to interpret the complex issues involved for those required to make policy decisions. Such specialized personnel need to be trained, hired and retained in service, so that they may serve their countries in these tasks.

The delegation of water resources management to the lowest appropriate level necessitates educating and training water management staff at all levels and ensuring that women participate equally in the education and training programmes. Particular emphasis has to be placed on the introduction of public participatory techniques, including enhancement of the role of women, youth, indigenous people and local communities. Skills related to various water management functions have to be developed by municipal government and water authorities, as well as in the private sector, local/national non-governmental organizations, cooperatives, corporations and other water-user groups. Education of the public regarding the importance of water and its proper management is also needed.

A pronounced need exists for novel training methods in integrated water resources management and planning, as well as to promote training on community participation approaches for water supply, sanitation and irrigation institutions. These concepts should be included as well in the training and education curricula for related professions in the water sector, specifically at the level of university and polytechnic education.

Because well-trained people are particularly important to water resources assessment and hydrological forecasting, personnel matters should receive special attention in this area. The aim should be to attract and retain personnel to work on water resources assessment who are sufficient in number and adequate in their level of education to ensure the effective implementation of the activities that are planned. Education may be called for at both the national and the international level, with adequate terms of employment being a national responsibility.

Innovative approaches should be adopted for professional and managerial staff training in order to cope with changing needs and challenges. Flexibility and adaptability regarding emerging water issues should be developed. Training activities should be undertaken periodically at all levels within the organizations and innovative teaching techniques adopted for specific issues including development of training skills, in-service training, problem-solving workshops and refresher training courses.

Countries must establish manpower development plans, taking into consideration present requirements and planned developments. Subsequently, the development and performance of country-level training institutions should be enhanced so that they can play a pivotal role in human resources development. It is also important that countries provide adequate training for women in the sustainable maintenance of equipment, water resources management and environmental sanitation.

Implicit in virtually all elements of this programme is the need for progressive enhancement of the training and career development of personnel at all levels in sector institutions. Specific programme activities will involve the training and retention of staff with skills in community involvement, low-cost technology, financial management, and integrated planning of urban water resources management.

Strategy and programme targets:

Human resources development should be undertaken by: (a) strengthening of training capacities within developing countries; (b) training of water managers at all levels so that they have an appropriate understanding of all the elements necessary for their decision making; (c) training of the necessary professionals, including extension workers; (d) training being given simultaneously at all levels; (e) improvement of career structures; (f) promotion of awareness-creation programmes, including mobilizing commitment and support at all levels and initiating global and local action to promote such programmes.

10

Mechanisms for implementation and coordination at global, national, regional and local levels

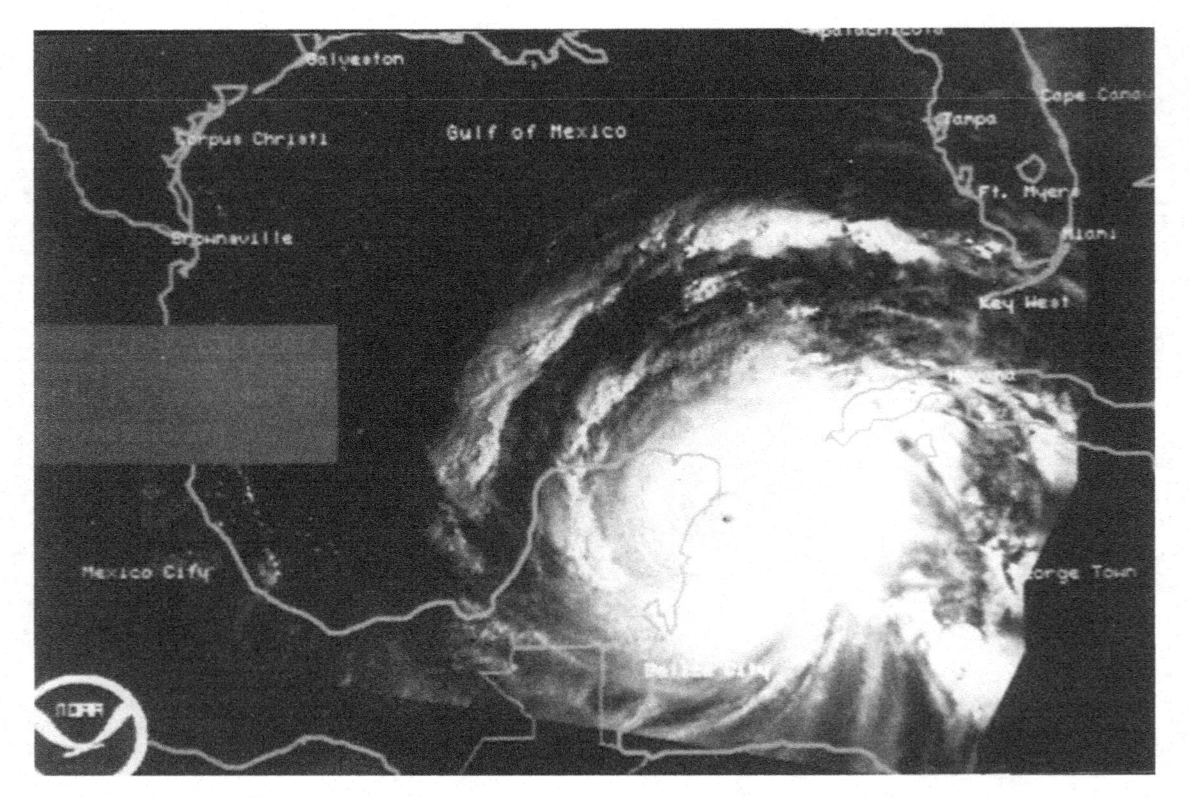

A satellite view of hurricane development in the Caribbean. This illustrates the need for international cooperation on water resource management. Credit: WMO.

10

Mechanisms for implementation and coordination at global, national, regional and local levels

10.1 Background

10.1.1 Appropriate levels of management

Centralized and sectoral (top down) approaches to water resources development and management have often proved inefficient or insufficient to solve local water management problems. The role of governments needs to change to ensure a more active participation of people and local institutions, public and private. The levels at which effective management decisions can be taken and problems can be solved vary from country to country, as do the relative roles of institutions in the administrative (socio-political) and hydrological (catchment/river basin) structures. The fundamental principle remains however that, in any given situation, water resources should be managed at the lowest appropriate levels. The need for integration of sustainable water management with land-use management, preservation of the environment and the reconciliation with other sectoral interests has to be recognized. In particular, the needs of human settlements, agriculture and industry have to be managed and based on a balanced consideration of the total needs of people and the environment.

10.1.2 Mechanisms at the national level

While the principle of the management of the resource at the lowest appropriate level requires a decentralized approach to water management, such an approach would fail if it were to operate in an institutional vacuum. There is a need for institutional arrangements at the national level, such as a national water authority, capable of defining priorities, policy directions, targets and, where appropriate, prescribing standards. The term 'authority' is used in this context to reflect its function of facilitating the implementation of water resources development and conservation activities and a system of checks and balances to safeguard public and national interests and to promote improved management.

The most pivotal and complex function of a national authority lies in the establishment of effective integration of the overall socio-economic and environmental decision-making process with the formulation of water resources policies and programmes. Similar linkages are required in order to conserve ecosystems and promote development needs on a sustainable basis. Such an authority may also provide an enabling environment for local resource mobilization and the flow of financial resources and the coordination of external support. Other functions of a national authority could be concerned with the coordination and management of data, including national monitoring networks, the formulation of a regulatory framework, the facilitation of technology transfer, the support of human resources development, the promotion of sustainable water management and full public participation in all aspects of water.

The national authority provides the necessary support for river and lake basin authorities or committees with responsibility for the integrated management of the water resources in the basin. At the very least, a central authority needs to provide a system of linkages between existing organizations dealing with water resources with a view to harmonizing approaches and policies. In the case of federated countries, parallel states or provincial authorities may be needed to perform related functions falling under the jurisdiction of states or provinces.

10.1.3 Transboundary basins

The most appropriate geographical entity for the planning and management of water resources is the basin, including its surface water and groundwater. Ideally, the effective integrated planning and development of transboundary river or lake basins has similar institutional requirements as in the case of a basin entirely within one country and should be based on the same principles. The essential function of existing international river basin organizations is one of reconciling and harmonizing the interests of riparian countries, monitoring water quantity and quality, development of concerted action programmes, exchange of information, and enforcing agreements. As regards transboundary groundwater basins, exploitation of such aquifers should take into account the safe aquifer yield, while developing the principles for the control of pollution.

10.1.4 International implementation mechanisms

Regional and global dimensions of water problems are rapidly growing in importance. Water resources are coming under increasing stress with the growth of population, while water is now recognized as the central component of global ecosystems and of the climate system. This requires the strengthening and development of a framework for the formulation of international water-related monitoring, programmes, policies and strategies. Such a framework must ensure that water problems are considered in the wider context of environmental and sustainable development issues.

10.1.5 Global coordination

At present, there is a lack of international arrangements for the effective coordination of global freshwater activities by multi-lateral, bilateral, and non-governmental organizations. A major problem is the lack of linkages between the community of external support agencies, governments and non-governmental organizations which deal with water resources coordination and facilitation.

Major recommendations proposed at the International Conference on Water and the Environment were that

> In the framework of the follow-up procedures developed by UNCED for Agenda 21, all Governments should initiate periodic assessments of progress. At the international level, United Nations institutions concerned with water should be strengthened to undertake the assessment and follow-up process. In addition, to involve private institutions, regional and non-governmental organizations along with all interested governments in the assessment and follow-up ... a world water forum or council [is proposed] to which all such groups could adhere. (Dublin Statement, Annex 1.)

10.2 Monitoring and information management

Basis for action:

Information on water resources and related socio-economic issues is essential to integrated water resources management. The basis for action has been elaborated in more detail in Chapter 3.

Strategy and programme targets:

Establish a monitoring and data network at national and international levels for the evaluation of the current situation and prediction of future problems.

10.3 Financing of water resources development and conservation

Basis for action:

There is a need to increase the flow of financial resources for water resources development and conservation.

Table 43. *Monitoring and information management*

Activities and related means of implementation	Level[1]	Considered by	
		ICWE[2]	UNCED[3]
1. Strengthening of networks for the assessment of surface and groundwater resources in terms of quantity and quality.	TNPL	X	X
2. Establishment of coordination mechanisms for water data collection and analysis and for integration with socio-economic data.	RTNP	X	X
3. Establishment of an integrated international network for the use of information in the monitoring of issues of an international nature.	GRTN	X	X
4. Support from the international community for the strengthening of national capabilities and for the establishment of an international network.	GR	X	X
5. Establishment of an international data clearing system for data management.	GRTN	X	X
6. Study of the feasibility of establishing a geographical information system or interactive systems for the storage, processing and analysis of information.	GRTNP	X	X
7. Assessment of water quantity and quality for policy and programme formulation.	GRTNPL	X	X

[1] Level of implementation: G=Global; R=Regional; T=Transboundary; N=National; P=Provincial; L=Local.
[2] Considered in ICWE *Report of the Conference* section 7.
[3] Considered in UNCED Agenda 21 throughout the document.

Strategy and programme targets:

To develop a system of credits and loan guarantees for direct financing of services, based on the principle of water as an economic good.

10.4 Development of national managerial/administrative infrastructures

Basis for action:

Efficient integrated management requires the devolution of managerial responsibilities to the lowest appropriate level and the existence of suitable management/administrative units at all relevant levels.

Table 44. *Financing of water resources development and conservation*

Activities and related means of implementation	Level[1]	Considered by	
		ICWE[2]	UNCED[3]
1. Ensure the financial and administrative autonomy of local service organizations.	TNPL	X	X
2. Ensure full cost recovery for local services including environmental costs, and provisions for reimbursing direct subsidies to users.	NPL	X	X
3. Support the development of financial facilities such as local revolving funds with support from regional, national and international financing institutions.	GR NPL	X	X
4. Establish or strengthen local, regional and national financial institutions to provide loans to service autonomous service organizations or to guarantee such loans from local sources.	NPL	X	X
5. Strengthen international support to national efforts to increase the flow of financing.	GR	X	X
6. Review, where necessary, rules and regulations of international financial institutions to allow the direct financing of services.	GR	X	X

[1] Level of implementation: G=Global; R=Regional; T=Transboundary; N=National; P=Provincial; L=Local.
[2] Considered in ICWE *Report of the Conference* section 7.
[3] Considered in UNCED Agenda 21 throughout the document.

Strategy and programme targets:

Establish or strengthen organizations at all appropriate levels for the integrated development and management of water resources.

10.5 Cooperation and coordination at the international level

Basis for action:

The effectiveness of interventions to deal with existing or emerging issues of global concern will often require the formulation of concerted policies and approaches to be carried out by countries individually or jointly, or by international organizations.

Strategy and programme targets:

Establish or strengthen informal and formal institutional mechanisms for cooperation and coordination at all relevant international levels.

Table 45. *Development of national managerial/administrative infrastructures*

Activities and related means of implementation	Level[1]	Considered by	
		ICWE[2]	UNCED[3]
1. Provide concerted support to national efforts to assess current organizational arrangements, to define institutional requirements and to rationalize integrated management procedures.	GR	X	X
2. Establish national water authorities to oversee the integrated development and conservation of water resources and to provide guidelines, standards and regulations.	N	X	X
3. Support and strengthen further development of water legislation and institutional mechanisms for the coordinated management of water resources to ensure sustainable development.	N	X	X
4. Ensure that effective flood and drought warning and preparedness systems are part of national sustainable development plans, within the framework of the International Decade for Natural Disaster Reduction.	N	X	X
5. Establish regional entities responsible for the integration of regional concerns into the management process.	NP	X	X
6. Establish river and lake basin authorities to have responsibility for the management of basins.	NP	X	X
7. Establish local authorities to monitor local conditions, to ensure the achievement of local objectives, and the participation of the community, in particular of women.	PL	X	X
8. Grant managerial and administrative autonomy to service organizations with proper governmental guidelines and controls. Support the establishment of user organizations.	NPL	X	X

[1] Level of implementation: G=Global; R=Regional; T=Transboundary; N=National; P=Provincial; L=Local.
[2] Considered in ICWE *Report of the Conference* section 7.
[3] Considered in UNCED Agenda 21 throughout the document.

Table 46. *Cooperation and coordination at the international level*

Activities and related means of implementation	Level[1]	Considered by	
		ICWE[2]	UNCED[3]
1. Promote better management within transboundary river and lake basins through: (a) evaluating the experience gained with existing transboundary basin water authorities, committees and commissions; (b) establishing transboundary river/lake basin organizations to manage in harmony the planning, development and conservation of transboundary resources, with well defined objectives agreed to by the riparian countries involved; (c) supporting the further development of legal principles and institutional mechanisms for the coordination of water management within transboundary basins; (d) promoting the cooperation of riparian countries within transboundary basins in the establishment of appropriate legal, institutional and operational mechanisms; (e) providing a forum to encourage dialogue and confidence-building measures regarding transboundary issues among riparian countries.	TN	X	X
2. Promote regional coordination through: (a) strengthening of existing regional mechanisms of a continental and sub-continental scale in order to harmonize policies, strategies and programmes; (b) promotion of cooperation on a regional basis for the exchange of experience in the protection and use of transboundary waters, including legal mechanisms and institutions.	GRTN	X	X
3. Promote information exchange through: (a) exchange of data on the components of the water cycle in terms of quantity and quality through established international programmes; (b) establishment and use of international data centres, supported by regulations under international law specifying mutual obligations and rules of procedure; (c) facilitating discussions and the development of recommendations, based on shared experience in areas related to water management such as legislation, research and development, and technology transfer.	GR N	X	X
4. Strengthen United Nations institutions such as the ECOSOC Committee on Natural Resources and enhance inter-agency coordination through mechanisms such as the Intersecretariat Group for Water Resources.	GR	X	X
5. Strengthen coordination between the United Nations system and other bilateral and multi-lateral governmental and non-governmental organizations through: (a) reviewing the functions and capacities of inter-governmental bodies, including United Nations institutions, regional and sub-regional organizations and non-governmental organizations; (b) identifying key needs not adequately addressed by these bodies; (c) making concrete proposals for strengthening inter-governmental bodies and further improving coordination with Member States.	GRTN	X	X
6. Develop concerted approaches to financial and technical cooperation among external support agencies and governments of developing countries, particularly with regard to the strengthening of national capacity.	GR N	X	X
7. Support the establishment of a facilitating mechanism, such as a world water forum or council, by which water specialists in various constituencies such as governments, international bodies, non-governmental organizations and private sector bodies could cooperate.	G N	X	X
8. Raise the global profile of freshwater issues and maintain that profile in the follow up to the United Nations Conference on Environment and Development, with a view to ensuring sustained global commitment to addressing freshwater problems.	G N	X	X

[1] Level of implementation: G=Global; R=Regional; T=Transboundary; N=National; P=Provincial; L=Local.
[2] Considered in ICWE *Report of the Conference* section 7.
[3] Considered in UNCED Agenda 21 throughout the document.

Table 47. *Development of a framework for the formulation of international policies, strategies and programmes*

Activities and related means of implementation	Level[1]	Considered by	
		ICWE[2]	UNCED[3]
1. Review functions of existing intergovernmental bodies dealing with water resources with a view to improving the monitoring of events and developing recommendations, guidelines, standards, and conventions.	GR N	X	X
2. Develop a charter of rights and duties of national and international organizations in the development, utilization and conservation of water resources through the appropriate intergovernmental body.	GR N	X	X
3. Review the existing capacity of secretariats of intergovernmental bodies to provide support to their respective bodies in a concerted manner.	GR N	X	X

[1] Level of implementation: G=Global; R=Regional; T=Transboundary; N=National; P=Provincial; L=Local.
[2] Considered in ICWE *Report of the Conference* section 7.
[3] Considered in UNCED Agenda 21 throughout the document.

10.6 Development of a framework for the formulation of international policies, strategies and programmes

Basis for action:

With the increasing importance of the regional and global dimensions of water problems, the ability to monitor events and predict emerging issues is increasing in importance. So is the need for concerted regional and global policies, strategies and programmes.

Strategy and programme targets:

To strengthen the regional and global intergovernmental capacity to formulate and implement policies, strategies and programmes.

Annex 1

The Dublin Statement on water and sustainable development

Scarcity and misuse of fresh water pose a serious and growing threat to sustainable development and protection of the environment. Human health and welfare, food security, industrial development and the ecosystems on which they depend, are all at risk, unless water and land resources are managed more effectively in the present decade and beyond than they have been in the past.

Five hundred participants, including government-designated experts from a hundred countries and representatives of eighty international, intergovernmental and non-governmental organizations attended the International Conference on Water and the Environment (ICWE) in Dublin, Ireland, on 26–31 January 1992. The experts saw the emerging global water resources picture as critical. At its closing session, the Conference adopted this Dublin Statement and the Conference Report. The problems highlighted are not speculative in nature; nor are they likely to affect our planet only in the distant future. They are here and they affect humanity now. The future survival of many millions of people demands immediate and effective action.

The Conference participants call for fundamental new approaches to the assessment, development and management of freshwater resources, which can only be brought about through political commitment and involvement from the highest levels of government to the smallest communities. Commitment will need to be backed by substantial and immediate investments, public awareness campaigns, legislative and institutional changes, technology development, and capacity building programmes. Underlying all these must be a greater recognition of the interdependence of all peoples, and of their place in the natural world.

In commending this Dublin Statement to the world leaders assembled at the United Nations Conference on Environment and Development (UNCED) in Rio de Janeiro in June 1992, the Conference participants urge all governments to study carefully the specific activities and means of implementation recommended in the Conference Report, and to translate those recommendations into urgent action programmes for water and sustainable development.

Guiding priniciples

Concerted action is needed to reverse the present trends of overconsumption, pollution, and rising threats from drought and floods. The Conference Report sets out recommendations for action at local, national and international levels, based on four guiding principles.

Principle No. 1

Fresh water is a finite and vulnerable resource, essential to sustain life, development and the environment

Since water sustains life, effective management of water resources demands a holistic approach, linking social and economic development with protection of natural ecosystems. Effective management links land and water uses across the whole of a catchment area or groundwater aquifer.

Principle No. 2

Water development and management should be based on a participatory approach, involving users, planners and policy-makers at all levels

The participatory approach involves raising awareness of the importance of water among policy-makers and the general public. It means that decisions are taken at the lowest appropriate level, with full public consultation and involvement of users in the planning and implementation of water projects.

Principle No. 3

Women play a central part in the provision, management and safeguarding of water

This pivotal role of women as providers and users of water and guardians of the living environment has seldom been reflected in institutional arrangements for the development and management of water resources. Acceptance and implementation of this principle requires positive policies to address women's specific needs and to equip and empower women to participate at all levels in water resources programmes, including decision-making and implementation, in ways defined by them.

Principle No. 4

Water has an economic value in all its competing uses and should be recognized as an economic good

Within this principle, it is vital to recognize first the basic right of all human beings to have access to clean water and sanitation at an affordable price. Past failure to recognize the economic value of water has led to wasteful and environmentally damaging uses of the resource. Managing water as an economic good is an important way of achieving efficient and equitable use, and of encouraging conservation and protection of water resources.

The action agenda

Based on these four guiding principles, the Conference participants developed recommendations which enable countries to tackle their water resources problems on a wide range of fronts. The major benefits to come from implementation of the Dublin recommendations will be:

Alleviation of poverty and disease

At the start of the 1990s, more than a quarter of the world's population still lack the basic human needs of enough food to eat, a clean water supply and hygienic means of sanitation. The Conference recommends that priority be given in water resources development and management to the accelerated provision of food, water and sanitation to these unserved millions.

Protection against natural disasters

Lack of preparedness, often aggravated by lack of data, means that droughts and floods take a huge toll in deaths, misery and economic loss. Economic losses from natural disasters, including floods and droughts, increased three-fold between the 1960s and the 1980s. Development is being set back for years in some developing countries, because investments have not been made in basic data collection and disaster preparedness. Projected climate change and rising sea-levels will intensify the risk for some, while also threatening the apparent security of existing water resources.

Damages and loss of life from floods and droughts can be drastically reduced by the disaster preparedness actions recommended in the Dublin Conference Report.

Water conservation and reuse

Current patterns of water use involve excessive waste. There is great scope for water savings in agriculture, in industry and in domestic water supplies.

Irrigated agriculture accounts for about 80% of water withdrawals in the world. In many irrigation schemes, up to 60% of this water is lost on its way from the source to the plant. More efficient irrigation practices will lead to substantial freshwater savings.

Recycling could reduce the consumption of many industrial consumers by 50% or more, with the additional benefit of reduced pollution. Application of the 'polluter pays' principle and realistic water pricing will encourage conservation and reuse. On average, 36% of the water produced by urban water utilities in developing countries is 'unaccounted for'. Better management could reduce these costly losses.

Combined savings in agriculture, industry and domestic water supplies could significantly defer investment in costly new water-resource development and have enormous impact on the sustainability of future supplies. More savings will come from multiple use of water. Compliance with effective discharge standards, based on new water protection objectives, will enable successive downstream consumers to reuse water which presently is too contaminated after the first use.

Sustainable urban development

The sustainability of urban growth is threatened by curtailment of the copious supplies of cheap water, as a result of the depletion and degradation caused by past profligacy. After a generation or more of excessive water use and reckless discharge of municipal and industrial wastes, the situation in the majority of the world's major cities is appalling and getting worse. As water scarcity and pollution force development of ever more distant sources, marginal costs of meeting fresh demands are growing rapidly. Future guaranteed supplies must be based on appropriate water charges and discharge controls. Residual contamination of land and water can no longer be seen as a reasonable trade-off for the jobs and prosperity brought by industrial growth.

Agricultural production and rural water supply

Achieving food security is a high priority in many countries, and agriculture must not only provide food for rising populations, but also save water for other uses. The challenge is to develop and apply water-saving technology and management methods, and, through capacity building, enable communities to introduce institutions and incentives for the rural population to adopt new approaches, for both rainfed and irrigated agriculture. The rural population must also have better access to a potable water supply and to sanitation services. It is an immense task, but not an impossible one – provided appropriate policies and programmes are adopted at all levels—local, national and international.

Protecting aquatic ecosystems

Water is a vital part of the environment and a home for many forms of life on which the well-being of humans ultimately depends. Disruption of flows has reduced the productivity of many such ecosystems, devastated fisheries, agriculture and grazing, and marginalized the rural communities which rely on these. Various kinds of pollution, including transboundary pollution, exacerbate these problems, degrade water supplies, require more expensive water treatment, destroy aquatic fauna, and deny recreation opportunities.

Integrated management of river basins provides the opportunity to safeguard aquatic ecosystems, and make their benefits available to society on a sustainable basis.

Resolving water conflicts

The most appropriate geographical entity for the planning and management of water resources is the river basin, including surface and groundwater. Ideally, the effective integrated planning and development of transboundary river or lake basins has similar institutional requirements to a basin entirely within one country. The essential function of existing international basin organizations is one of reconciling and harmonizing the interests of riparian countries, monitoring water quantity and quality, development of concerted action programmes, exchange of information, and enforcing agreements.

In the coming decades, management of international watersheds will greatly increase in importance. A high priority should therefore be given to the preparation and implementation of integrated management plans, endorsed by all affected governments and backed by international agreements.

The enabling environment

Implementation of action programmes for water and sustainable development will require a substantial investment, not only in the capital projects concerned, but, crucially, in building the capacity of people and institutions to plan and implement those projects.

The knowledge base

Measurement of components of the water cycle, in quantity and quality, and of other characteristics of the environment affecting water are an essential basis for undertaking effective water management. Research and analysis techniques, applied on an interdisciplinary basis, permit the understanding of these data and their application to many uses.

With the threat of global warming due to increasing greenhouse gas concentrations in the atmosphere, the need for measurements and data exchange on the hydrological cycle on a global scale is evident. The data are required to understand both the world's climate system and the potential impacts on water resources of climate change and sea level rise. All countries must participate and, where necessary, be assisted to take part in the global monitoring, the study of the effects and the development of appropriate response strategies.

Capacity building

All actions identified in the Dublin Conference Report require well-trained and qualified personnel. Countries should identify, as part of national development plans, training needs for water-resources assessment and management, and take steps internally and, if necessary with technical co-operation agencies, to provide the required training, and working conditions which help to retain the trained personnel.

Governments must also assess their capacity to equip their water and other specialists to implement the full range of activities for integrated water-resources management. This requires provision of an enabling environment in terms of institutional and legal arrangements, including those for effective water-demand management.

Awareness raising is a vital part of a participatory approach to water resources management. Information, education and communication support programmes must be an integral part of the development process.

Follow-up

Experience has shown that progress towards implementing the actions and achieving the goals of water programmes requires follow-up mechanisms for periodic assessments at national and international levels.

In the framework of the follow-up procedures developed by UNCED for Agenda 21, all Governments should initiate periodic assessments of progress. At the international level, United Nations institutions concerned with water should be strengthened to undertake the assessment and follow-up process. In addition, to involve private institutions, regional and non-governmental organizations along with all interested governments in the assessment and follow-up, the Conference proposes, for consideration by UNCED, a world water forum or council to which all such groups could adhere.

It is proposed that the first full assessment on implementation of the recommended programme should be undertaken by the year 2000.

UNCED is urged to consider the financial requirements for water-related programmes, in accordance with the above principles, in the funding for implementation of Agenda 21. Such considerations must include realistic targets for the timeframe for implementation of the programmes, the internal and external resources needed, and the means of mobilizing these.

The International Conference on Water and the Environment began with a Water Ceremony in which children from all parts of the world made a moving plea to the assembled experts to play their part in preserving precious water resources for future generations.

In transmitting this Dublin Statement to a world audience, the Conference participants urge all those involved in the development and management of our water resources to allow the message of those children to direct their future actions.

Annex 2

Targets and cost estimates

For the purpose of developing the UNCED Agenda 21 chapter on protection of the quality and supply of freshwater resources, the Inter-Secretariat Group on Water Resources undertook to identify targets and to prepare cost estimates related to these targets. These estimates are based upon past and present experience and project annual average costs for the remainder of the century. Total investment needs and the portions required through external support were quantified to the extent possible.

These are indicative and order of magnitude estimates only and have not been reviewed by Governments. Actual costs and financial terms, including any that are non-concessional, will depend upon, inter alia, the specific strategies and programmes Governments decide upon for implementation.

The targets and cost estimates for the key programme areas within the freshwater sector are as follows:

Integrated water resources development and management

(i) Targets

(a) All countries, in accordance with their capacities and resources available, should have designed and initiated costed and targeted national action programmes, and 75% of all countries should have appropriate institutional structures and legal instruments in place by the year 2000;
(b) Demand management should be introduced into all national action plans and implemented by the year 2000; the necessary training and transfer of technology should have taken place and at least half the developing countries should have carried out evaluations on the effectiveness of demand management;
(c) Self-evaluations by countries should be performed in the year 2000 to measure qualitatively the extent to which public participation has been enhanced and its impact on programme effectiveness;
(d) Sub-sectoral targets of all freshwater programme areas should be reached by the year 2000.

(ii) Cost estimates

During the period 1993 to 2000, an annual amount of about US$ 100 million of international financing is required to support national development in this programme area. The strengthening of international institutions in support of the planning and initiation phases at the country level requires the allocation of about US$ 10 million per year. Transboundary and global freshwater issues require a financial support in the order of US$ 5 million annually for the executing national, regional and global authorities and organizations. The total annual financing requirements in this programme area amount to about US$ 115 million*.

Water resources assessment

(i) Targets

(a) All countries, appropriate to their individual capacities and available resources, should have studied in detail the feasibility of installing water resources assessment services by the year 2000;
(b) There should be water resources assessment services with a high-density hydrometric network installed in 70 countries, and services with limited but adequate capacity in 60 countries by the year 2000;
(c) There should be 110 countries with fully developed services, and 40 countries with services of a limited capacity by the year 2025;
(d) The longer-term target is to have fully operational services, based upon high-density hydrometric networks, available in all countries, appropriate to their individual capacities and available resources.

(ii) Cost estimates

In order to attain the targets for the year 2000, total average annual funding in the order of US$ 355 million is required, including contributions from external sources of US$ 145 million annually. The strengthening of international institutions for the development and exchange of information and technology requires about US$ 5 million per year. The total international financing needs for the years 1993 to 2000 for this programme area amount to US$ 145 million annually†.

Protection of water resources, water quality and aquatic ecosystems

(i) Targets

(a) All countries, appropriate to their capacities and resources available, should have identified all potential sources of water supply and prepared outlines for their protection, conservation and rational use by the year 2000;
(b) All countries should have effective water pollution control programmes, defined as enforceable standards for major point-source discharges and high-risk non-point sources, commensurate with their socio-economic development by the year 2000;

* When estimating the financial requirements for integrated water resources management it has to be kept in mind that the bulk of investments and external donor support is covered under the programme areas on protection of water resources, urban water management and rural waste management.
† Financial support is required primarily for the establishment and strengthening of national hydrometric networks, also for those covering transboundary watercourses.

(c) All countries should participate, as far as appropriate, in inter-country and international water quality monitoring and management programs such as Global Water Quality Monitoring Programme GEMS/WATER, UNEP's Environmentally Sound Management of Inland Waters, FAO's regional inland fishery bodies, and the RAMSAR Convention on Wetlands of International Importance especially as Waterfowl Habitat by the year 2000;
(d) The prevalence of water-associated diseases should be drastically reduced by the year 2025, starting with the eradication of dracunculosis (Guinea worm) and onchocerciasis (river blindness) by the year 2000;
(e) All countries should have established biological, health, physical and chemical quality criteria for all water bodies (surface and groundwater) with a view to an ongoing improvement of water quality by the year 2025.

(ii) Cost estimates

Funds for pollution control have to be generated ultimately within each country or river basin through cost recovery and economic or fiscal instruments. The "polluter pays principle" (PPP) has to be adopted in conformity with the notion of water as an economic good. Total costs, including those financed nationally, are estimated at about US$ 1.0 billion annually for the period 1993 to 2000. Of this amount about US$ 330 million would be needed annually from international sources*.

The assessment of global environmental issues also includes water quality and aquatic ecosystems monitoring and assessment. River monitoring for global flux estimates is covered under water resources assessment already, and the necessary funds indicated there. About US$ 10 million annually would be needed in addition to this to strengthen international institutions. The total financing requirement for this programme area amounts to US$ 340 million annually†.

Impacts of climate change on water resources

(i) Targets

(a) To understand and quantify the threat of the impact of climate change on freshwater resources by the year 2000;
(b) To facilitate the implementation of effective national counter-measures, as and when the threatening impact is seen as sufficiently confirmed to justify such action.

(ii) Cost estimates

During the years 1993 to 2000, national programmes to assess and mitigate the effects of climate change will incur costs of about US$ 100 million annually, of which about US$ 40 million would have to come from international sources‡.

* These costs are estimated for the establishment of legal, regulatory, institutional and technical measures to control water pollution.

† Costs for erosion control, sewage treatment and infrastructure measures are contained in relevant other chapters of Agenda 21.

‡ The impacts of climate change on freshwaters are costed with regard to sea level rise, droughts and increased flood risks.

Water and sustainable urban development

(i) Targets

(a) Medium- and long-term plans for environmental sanitation should be established by 1995 to ensure permanent protection of vulnerable groups against disease risks, especially cholera;
(b) A water resources master plan matching development objectives and enabling cities to plan on the basis of assured water allocation should be developed by the year 2000;
(c) Water and effluent quality standards should be enforced reflecting the polluter pays principle by the year 2000;
(d) Programmes to provide sanitary containment or treatment for at least 50% of the pollution load (in terms of biological oxygen demand) from domestic wastes should be initiated by 2002, and every country should have achieved river water quality (varying from location to location) which safeguards supplies for downstream users by the year 2015;
(e) In urban areas, by the year 2000, the number of people lacking water supply and sanitation services in 1990 should be halved, with full coverage by 2015, and a minimum of 85% of installed facilities should function at any given time and water be made available for 24 hours a day.

(ii) Cost estimates

It is estimated that the average total annual cost (1993–2000) of implementing the activities of this programme would be about $20 billion, including about $4.5 billion from the international community on grant or concessional terms*.

Water for sustainable food production and rural development

(i) Targets

(a) To achieve the improvement and modernization of 12 million hectares of irrigated land by the year 2000;
(b) To establish small-scale water programmes to improve a total of 10 million hectares of rainfed arable land by the year 2000;
(c) To achieve the expansion of irrigated areas by 21 million hectares, where environmentally sound, by the year 2000;
(d) To provide improved drainage and reduction of excessive losses to groundwater on 7 million hectares of irrigated land by the year 2000;
(e) To increase capture fishery production from 7 million to 10 million tons/year, and inland aquaculture production from 7 million to 14 millions tons/year by the year 2000.

* Cost estimates for drinking water supply and basic sanitation in rural areas are already included under drinking water supply and sanitation.

(ii) Cost estimates

It is estimated that the average total annual cost (1993–2000) of implementing the activities of this programme would be about $13.2 billion, including about $4.5 billion from the international community on grant or concessional terms*.

Drinking water supply and sanitation

(i) Targets

(a) The coverage of community water supply and sanitation services should be expanded to an additional 1.6 billion people by the year 2000;
(b) Full coverage in water supply should be achieved by the year 2025.

(ii) Cost estimates

Accelerated development is necessary to reach the desired coverage of water supply and basic sanitation services by the year 2000. The rate of investment for the years until 2000 would have to be doubled to a total of US$ 20 billion annually if complete service coverage were to be reached. The external component should be maintained at no less than one third of this, i.e. at about US$ 6.7 billion annually. The improved operation, maintenance and management of systems and the full utilization of the investments made requires the allocation of external support in the order of US$ 0.7 billion. The total external funding needs until the year 2000 are, therefore, US$ 7.4 billion annually†.

Summary of funding needs, per year 1993–2000

Programme area	Total annual budget (US$m)	From external sources (US$m)
Integrated water resources development and management	115	115
Water resources assessment	355	145
Protection of water resources, water quality and aquatic ecosystems	1 000	340
Impacts of climate change on water resources	100	44
Water and sustainable urban development	20 000	4 500
Water for sustainable food production and rural development	13 200	4 500
Drinking water supply and sanitation	20 000	7 400
Total Programme	54 770	17 040

* The cost estimates for water-based agricultural development activities include irrigation schemes, rainfed agriculture, livestock supply, inland fisheries and aquaculture in a total of 130 developing countries.

† The New Delhi Statement concludes that "if costs were halved and financial resources at least doubled, universal coverage could be within range by the end of the century". (See also United Nations General Assembly resolution A/RES/45/181.)

Annex 3

Participants at the International Conference on Water and the Environment

1. Government-designated experts

Algeria
Ait-Amara, Mr A.
Kherraz, Mr K.
Figuerero, Mr J.M.
Argentina
Fuschini-Mejia, Mr M.C.
Solar Dorrego, Mr L.
Australia
Constable, Mr D.J.
Filipetto, Ms L.
Ludlow, Ms J.
McCarthy, Mr T.
Roberts, Mr T.
Stewart, Mr B.J.
Austria
Grath, Mr J.
Nobilis, Mr F.
Bahamas
Weech, Mr P.
Bahrain
Ali Abdulla, Mr P.
Bangladesh
Ahmed, Mr R.
Hannan, Mr A.
Islam, Mr M.A.
Nishat, Mr A.
Belgium
De Brabander, Mr K.
Van Der Beken, Mr A.
Benin
Abouki, Mr M.
Bolivia
Salas, Mr R.E.
Botswana
Sekwale, Mr M.
Brazil
Daniel, Mr M.C.M.
Proenca Rosa, Mr C.A.
Ricarte, Mr A.O.S.
Bulgaria
Mandadjiev, Mr D.
Burundi
Ntahuga, Mr L.
Cambodia
Koum, Mr S.
Cameroon
Nkoulou Ntere, Mr P.
Canada
Bezeredi, Ms A.
Bruce, Mr J.P.
Davis, Mr D.A.
Grover, Mr B.
Hill, Mr H.
McRae, Mr T.
Cape Verde
Vieira, Mr H.J.
Central African Republic
Feizoure, Mr C.T.
Chad
Alainaye, Mr D.

Chile
Berguno, Mr B.
Manriquez Lobos, Mr G.
Sanchez, Mr V.
China
Wang Weizhong, Mr
Yan Hongbang, Mr
Yang Dingyuan, Mr
Ye Yongyi, Mr
Zheng Rugang, Mr
Colombia
Barros Luque, Mr R.A.
Melendez, Mr R.
Ramirez Vallejo, Mr J.
Congo
Goma, Mr Ph.
Costa Rica
Calvo Zeledon, Mr R.
Côte d'Ivoire
Kakadie, Mr Y.G.
Sakho, Mr M.A.
Cuba
Arrue Avila, Mr A.
Czechoslovakia
Kazimour, Mr V.
Kinkor, Mr J.
Molnar, Mr L.
Zavadsky, Mr I.
Democratic Rep. of Korea
Hong Yong, Mr
Pook Chon Sok, Mr
Denmark
Boesen, Mr J.
Jonch-Clausen, Mr T.
Korkman, Mr T.E.
Refsgaard, Mr J.C.
Storgaard Madsen, Ms B.
Ecuador
Rodriguez, Mr L.
Egypt
Abou El Dahab, Mr M.
Abu Zeid, Mr M.
Gamil, Mr E.M.
Mahmoud, Mr G.
Mesharafa, Mr H.
Moussa, Ms S.
Moustafa, Mr A.T.
Raafat, Mr F.
Ethiopia
Dejene, Mr W.M.
Seyoum, Mr H.S.
Tsegay, Mr A.
Finland
Haunia, Ms S.
Haverinen, Mr A.
Kontula, Mr E.
Nyroos, Ms H.
France
Charbonnel, Mr L.
Geny, Mr P.
Jaouen, Ms A.
Le Masson, Mr H.
Roussel, Ms O.
Truchot, Mr C.
Wagner, Mr M.
Gabon
Maganga-Nziengui, Mr A.
Otchanga, Mr W.
Gambia
Sahor, Mr M.
Samba, Mr S.
Germany
Erbel, Mr K.
Hofius, Mr K.
Roser, Mrs. S.
Rudolf, Mr B.
Teuber, Mr W.
Walch, Mr H-J.
Winzek, Mr H.
Ghana
Ayibotele, Mr N.B.
Greece
Denaxas, Mr E.
Karakatsoulis, Mr P.
Kolla-Mimikou, Ms M.
Megremis, Mr P.
Guinea
Diallo, Mr M. A.
Guinea-Bissau
Balde, Mr J. M.
Cardoso, Mr J.G.
Guyana
Pompey, Ms A.
Honduras
Burgos de Flores, Ms L.

Hungary
Hollo, Mr G.
Kisgyorgy, Mr S.
Nemeth, Mr M.
Ottlik, Mr P.
Starosolszky, Mr O.
Svetnik, Mr A.
India
Gupta, Mr D.B.
Kashyap, Mr R.
Tiwari, Mr D.S.
Venugopalan Nair, Mr J.
Indonesia
Alirahman, Mr
Soenarno, Mr
Iran, Islamic Republic of
Asgari, Mr A.
Jahani, Mr A.
Mahini, Mr S.S.
Mahmoudian, Mr S. A.
Massoumi-Alamouti, Mr A.
Youssefi-Zadeh, Mr M.
Ireland
Callan, Mr N.
Clarke, Mr B.
Dollard, Mr R.
Dooge , Mr J.C.I.
Moylan, Ms M.
McCumiskey, Mr B.
Israel
Ben-Zvi, Mr A.
Kahana, Mr Y.
Kantor, Mr S.
Sharma, Mr P.C.
Italy
Barni, Mr E.
Gigliani, Mr F.
Imparato, Mr I.G.
Moschetta, Mr G.
Olivieri, Ms V.
Romano, Mr E.
Scaroni, Mr A.
Tozzoli, Mr G.
Villa, Mr L.
Jamaica
Hardware, Mr T.W.
Japan
Nishimura, Mr Y.
Obayashi, Mr T.
Tsutsui, Mr H.
Yatsu, Mr R.
Yokouchi, Mr H.
Jordan
Qunqar, Mr E.
Kenya
Mwongera, Mr E.K.
Kuwait
Al Minayes, Mr A.M.
Al-Farhoud, Mr K.
Lao People's Dem. Rep.
Symmavong, Mr N.
Lebanon
Jaber, Mr B.
Rabbath, Mr A.
Lesotho
Makhoalibe, Mr S.
Liberia
Kroma, Mr A.
Malawi
Laisi, Mr E.Z.
Malaysia
Rosmah, Ms M.J.
Shahrizaila, Mr A.
Malta
De Ketelaere, Mr D.
Spiteri, Ms A.
Mauritania
Baba, Mr O.S.A.
Ould Dahi, Mr M.
Mauritius
Sok Appadu, Mr S.
Mexico
Calderón Bartheneuf, Mr J.
David, Mr A.
Espino de la O, Mr E.
Garduño Velasco, Mr H.
Glender, Mr A.
Romero Alvarez, Mr H.
Mongolia
Myagmarjav, Mr B.
Morocco
Hajji, Mr A.
Jellali, Mr M.
Mozambique
Cambula, Mr P.F.F.
Myanmar
U Tin Myint
Nepal
Sharma, Mr C.
Netherlands
Alaerts, Mr G.J.
Ardon, Mr W.G.
Blom, Mr J.
Koudstaal, Mr R.

Oudshoorn, Mr H.M.
Rijsberman, Mr F.
Savenije, Mr H.
Zijlmans, Mr R.
Zuidema, Mr F.C.

Netherlands Antilles
Newton, Mr E.C.
Statia, Mr T.B.

New Zealand
Mosley, Mr P.

Nicaragua
Gutierrez, Mr C.

Niger
Bako, Mr Y.

Nigeria
Abatcha, Mr A.A.
Aina, Mr E.O.A.
Bassey, Mr J.O.
Ettu, Mr S.A.
Hanidu, Mr J.A.
Imevbore, Mr A.M.A.
Okeke, Ms E.
Shaib, Mr B.
Udoeka, Mr E.D.
Wadibia-Anyanwu, Ms N.

Norway
Bendiksen, Ms R.
Eidheim, Ms I.
Hansen, Mr S.
McNeill, Mr D.
Tollan, Mr A.
Wangen, Mr G.

Oman
Al-Harthy, Mr S.S.
Al-Said, Mr B.
Al-Shaikh, Mr J.
Al-Shaqsi, Mr S.R.

Pakistan
Qaiser, Mr G.

Panama
Candanedo, Ms C.

Papua New Guinea
Douglas, Mr J.

Paraguay
Fragano , Mr. F.
Sanchez Guffanti, Mr G.

Peru
Ventura Napa, Mr M.

Philippines
Sosa, Mr L.

Poland
Kindler, Mr J.
Zielinski, Mr J.

Portugal
Almiro Do Vale, Mr F.
Bastos, Mr J.P.
Borrego, Mr C.
Candido, Mr A.
Cavalo, Mr A.
Espirito Santo, Ms F.
Gouveia, Ms T.
Lemos, Mr P.
Mendes, Mr A.
Passaro, Mr M.C.
Pires, Mr A.
Ramos, Ms L.

Romania
Serban, Mr P.

Sao Tome & Principe
Da Conceicao, Mr J.

Saudi Arabia
Abdulrazzak, Mr M.
Al Haratani, Mr E.
Al Kaltham, Mr M.
Al-Azzaz, Mr A.
Al-Saati, Mr A.J.
Al-Sahli, Mr M.J.
Almaziad, Mr A.
Haddad, Mr A.

Senegal
Fall, Mr C.
Sylla, Mr D.C.

Seychelles
Mascarenhas, Mr J.P.

Singapore
Ong, Mr H.S.
Tan, Mr L.

Spain
Fernandez, Mr M.A.
Mingo Magro, Mr J.

Sri Lanka
Wijesinghe, Mr M.W.P.

St. Vincent & the Grenadines
Cummings, Mr D.

Sudan
Elhag, Mr M.E.E.
Hidaytalla, Mr A.
Mohamed, Mr T.A.
Nour, Mr M.E.M.

Sweden
Andersson, Mr I.
Bjorklund, Ms G.
Falkenmark, Ms M.

Switzerland
Emmenegger, Mr C.
Flury, Mr M.

Goetz, Mr A.
Lazzarotto, Mr S.
Musy, Mr A.
Spreafico, Mr M.

Syrian Arab Rep.
Hadid, Mr B.

Thailand
Buddhapalit, Mr A.
Chindasanguan, Mr C.
Hungspreug, Mr S.

Tunisia
Horchani, Mr A.

Turkey
Bozkurt, Mr S.
Kuleli, Ms S.
Kulga, Mr D.
Numanoglu, Ms N.
Solen, Mr A.

Tuvalu
Sakaio, Mr V.P.

Uganda
Bomukama, Mr S.
Kagimu, Mr G.M.
Kahangire, Mr P.
Odul, Mr J.

United Kingdom
Cocking, Ms J.
Frampton, Mr R.
Kirby, Ms C.
Parks, Ms Y.
Pike, Mr T.
Rodda, Mr D.
Sherriff, Mr J.
Simcock, Mr A.
Wilkinson, Mr W.

United Republic of Tanzania
Msuya, Mr M.O.

United States of America
Austin, Mr J.
Dickey, Mr G.
Moody, Mr D.
Randall, Mr B.
Rogers, Mr P.
Schifferdecker, Mr A.
Stallings, Mr E.
Steever, Mr Z.
Walker, Mr C.
Wilcher, Ms L.

Uruguay
Arduino, Mr G.
Graceras, Mr C.
Rodriguez, Mr M.
Serrentino, Mr C.

Venezuela
Gonzalez, Mr C.

Viet Nam
Vu Van Tuan, Mr

Yemen
Al-Fusail, Mr A.K.
Mohamed, Mr N.

Yugoslavia
Radojicic, Mr L.

Zambia
Mbewe, Mr J.J.
Mbumwae, Mr L.L.

2. United Nations Agencies

Economic and Social Commission for Western Asia (ESCWA)

Radjai, Mr A.

Food and Agriculture Organization of the United Nations (FAO)

Burchi, Mr S.
De Haen, Mr H.
Kandiah, Mr A.
Kapetsky, Mr J.
Rubery, Mr N.
Rydzewski, Mr J.
Saouma, Mr E.
Scott, Mr S.
Sombroek, Mr A.
Stringer, Mr R.
Toros, Mr H.

International Research and Training Institute for the Advancement of Women (INSTRAW)

Bulajich, Ms B.
Shields, Ms M.

International Atomic Energy Agency (IAEA)

Crijns, Mr M.J.
Yurtsever, Mr Y.

Non-Governmental Liaison Service (NGLS)

Rodda, Ms A.

Organization of African Unity (OAU)

Diallo, Mr I. K.

PROWWESS, Afrique/United Nations Development Fund for Women (UNIFEM)

Traore, Ms A.

United National Center for Human Settlement (UNCHS HABITAT)

Dzikus, Mr A.
Ramachandran, Mr A.
Sinnatamby, Mr G.
Swan, Mr P.

United Nations Children's Fund (UNICEF)

De Rooy, Mr C.
Glattbach, Mr J.
Jolly, Mr R.
McLoughney, Mr E.
Rosenhall, Mr L

United Nations Conference on Environment and Development (UNCED)

Helmer, Mr R.
Steady, Ms F.C.
Wheeler, Mr J.

United Nations Dept. of Int. Economic and Social Affairs (UNDIESA)

Najlis, Mr P.

United Nations Department of Technical Co-operation for Development (UNDTCD)

Appleton, Mr B.
Edwards, Mr K.A.
Pastizzi-Ferencic, Ms D.
Ling Maung San, Mr
Sauveplane, Mr C.
Solanes, Mr M.
Vlachos, Mr E.

United Nations Development Programme (UNDP)

De Gala, Ms M.
Hartvelt, Mr F.
Helland-Hansen, Mr E.
Kakonge, Mr J.
Lowes, Mr P.
Okun, Mr D.
Rajeswary, Ms I.

United Nations Disaster Relief Co-ordinator, Office of the (UNDRO) and International Decade for Natural Disaster Reduction (IDNDR)

Nemec, Mr J.

United Nations Educational, Scientific and Cultural Organization (UNESCO)

Aureli, Ms A.
Bastide, Ms M.
Dumitrescu, Mr S.
Gladwell, Mr J.
Schetselaar, Mr E.M.
Szöllösi-Nagy, Mr A.
Tatit Holtz, Mr A.

United Nations Environment Programme (UNEP)

Biswas, Mr A.
El-Habr, Mr H.
Golubev, Mr G.
Illueca, Mr J.
Mageed, Mr Y.A.
Tolba, Mr M.
Vandeweerd, Ms V.
White, Mr G.

United Nations Institute for Training and Research (UNITAR)

Chossudovsky, Mr E.M.

United Nations Research Institute for Social Development (UNRISD)

Jha, Ms V.

United Nations Information Centre (UNIC)

De Wette, Mr J.

United Nations Sudano-Sahelian Office (UNSO)

Oerum, Mr T.

Water Supply and Sanitation Collaborative Council (WSS)

Catley-Carlson, Ms M.
Locke, Mr B.
Van Damme, Mr H.
Wirasinha, Mr R.

World Bank (WB)

Briscoe, Mr J.
Delli-Priscoli, Mr J.
Elahi, Mr A.M.
Evans, Mr T.
Feder, Mr G.
Garn, Mr M.H.
Kuffner, Mr U.
Le Moigne, Mr G.
Matthews, Mr G.
Monosowski, Ms E.
Rotival, Mr A.

World Health Organization (WHO)

Fenger, Mr B.
Fraser, Mr A.
Meybeck, Mr M.
Warner, Mr D.
Wong, Mr P.

World Meteorological Organization (WMO)

Askew, Mr A.
Burns, Ms M.
Dar-Ziv, Ms E.
Dengo, Mr M.
Espejo, Ms C.
Kraemer, Mr D.
Melder, Mr O-M.
Obasi, Mr G.O.P.
Pieyns, Mr S.
Rodda, Mr J.

3. Organizations with observer status with the United Nations

Abu-Gharbiyeh, Mr M.	Palestine

4. Representatives from Intergovernmental Organizations

Agence de Coopération Culturelle et Technique (ACCT)	Burton, Mr J.
Arab Centre for the Studies of Arid Zones (ACSAD)	Khouri, Mr J.
Asian Development Bank (ADB)	McIntosh, Mr A. Ch.
Comité Inter-Africain de Lutte contre la Sécheresse au Sahel (CILSS)	Ousmane, Mr B.
Comité Interafricain d'Etudes Hydrauliques (CIEH)	Diagana, Mr B.
Commission of the European Communities (CEC)	Mandl, Mr V.
	Clarke, Mr T.
	Piavaux, Mr A.
HYDROMET	Tawfik, Mr M.M.
International Institute for Applied Systems Analysis (IIASA)	Kulshrestha, Mr S.
	Somlyody, Mr L.
International Irrigation Management Institute (IIMI)	Lenton, Mr R.
Lake Chad Basin Commission (LCBC)	Irivboje, Mr O.C.
	Jauro, Mr A.B.
North Atlantic Treaty Organization (NATO)	Da Cunha, Mr L.V.
Organization for Economic Co-operation and Development (OECD)	Baile, Ms S.
Permanent Joint Technical Commission for the Nile (PJTC)	Ezzat, Mr M.N.
	Hamad, Mr B.M.
	Mohamed, Mr K.A.
	Seoud, Mr A.A.
Southern African Development Co-ordination Conference (SADCC)	Makhoalibe, Mr S.

5. Representatives from non-governmental organizations

American Water Works Association (AWWA) and Water for People	Nagle, Mr W.J.
Associacao Brasileira de Recursos Hidricos (ABRH)	Braga, Mr B.P. Canedo, Mr P.
Brazilian Committee for UNEP	Mattos De Lemos, Mr H.
CAPE '92	Parcells, Mr S.
Centre international pour la Formation à la Gestion des Ressources en Eau (CEFIGRE)	Robert, Mr D.J.
Committee on Water Research (COWAR)	Colenbrander, Mr H. Plate, Mr E.
Environmental Defense Fund (EDF)	Moore, Ms D.
Friends of the Earth/Earthwatch	Hamilton, Mr A.G.
Global Water Summit Initiative	Starr, Ms J.
Globetree Foundation	Van Bronckhorst, Mr B.
Greenpeace Ireland	Kinghan, Ms H.
Institut Méditerranéen de l'Eau (IME)	Margat, Mr J. Potie, Mr L.
Institution of Water and Environmental Management (IWEM)	Clarke, Mr K.
International Association on Water Pollution Research and Control (IAWPRC)	Milburn, Mr A.
International Association for Hydraulic Research (IAHR)	Muller, Mr A.
International Association for Water Law (IAWL)	Caponera, Mr D.
International Association of Hydrogeologists (IAH)	Llamas, Mr M.R. Skinner, Mr A.C.
International Association of Hydrological Sciences (IAHS)	Shamir, Mr U.
International Association of Theoretical and Applied Limnology (IAL)	Murray, Mr D.
International Commission of Water Law (ICWL)	Scannell, Ms Y.
International Commission on Irrigation and Drainage (ICID)	Hennessy, Mr J.
International Council of Environmental Law (ICEL)	Jorgensen, Mr S. McKeague, Ms P.
International Fertilizer Industry Association (IFA)	Stafford, Mr L.
International Lake Environment Committee Foundation (ILEC)	Moriya, Mr M.
International Life Sciences Institute (ILSI)	Julkunen, Ms P.
International Union for the Conservation of Nature (IUCN)	Dugan, Mr P.
International Union of Food Science and Technology (IUFOST)	Hood, Mr D.E.
International Water Resources Association (IWRA)	Stout, Mr G.
International Water Supply Association (IWSA)	Bays, Mr L. Tessendorf, Mr H.
International Water Tribunal (IWT)	Nollkaemper, Mr A.
RAMSAR Convention Bureau	Lethier, Mr H.
International Secretariat for Water (ISW)	Chabert d'Hières, Mr L. Jost, Mr R.
WaterAid	King, Mr N.
Water, Engineering and Development Centre (WEDC)	Franceys, Mr R.W.A.
Worldwide Fund for Nature (WWF)	King-Volcy, Ms N.

6. ICWE Secretariat

Anukam, Mr L.
Blanc, Ms V.
Diawara, Mr A.
Gorre-Dale, Ms E.
Ibrekk, Mr H.O.
Yabi, Ms M.
Young, Mr G.

7. DOE Ireland

Costigan, Ms M
Downes, Ms E.
Dunne, Mr D.
Dunne, Ms C.
Glynn, Ms A.
Keenan, Ms J.
Macken, Mr P.
McGuiness, Ms A.
Noone, Ms M.
Ryan, Mr C.

Annex 4

Financial contributors to the International Conference on Water and the Environment

Australia
Canada
Commission of the European Communities
Finland
France (ORSTOM)
Germany
Ireland
Netherlands
Norway
Sweden
Switzerland
United Kingdom of Great Britain and Northern Ireland
United States of America
United Nations Department of Technical Co-operation for Development (UN-DTCD)
United Nations Children's Fund (UNICEF)
United Nations Development Programme (UNDP)
United Nations Environment Programme (UNEP)
United Nations Centre for Human Settlements (HABITAT)
Food and Agriculture Organization (FAO)
United Nations Educational Scientific and Cultural Organization (UNESCO)
World Health Organization (WHO)
World Bank (IBRD)
World Meteorological Organization (WMO)
International Atomic Energy Agency (IAEA)
United Nations Conference on Environment and Development (UNCED)
British Hydrological Society
International Council of Scientific Unions (ICSU)
International Association of Hydrological Sciences (IAHS)

Annex 5

Members of the Steering Committee for the International Conference on Water and the Environment

John C. **RODDA**	Chairman, ACC ISGWR, (WMO)
Gordon J. **YOUNG**	Conference Co-ordinator, ICWE
Eirah **GORRE-DALE**	Public Information and Promotion Co-ordinator, ICWE(UNDP)
Lawrence **ANUKAM**	UNCED Liaison, NGO and IGO Affairs
Clare **DUNNE**	Department of the Environment (DOE), Ireland
Arthur **ASKEW**	WMO
Borjana **BULAJICH**	UN-INSTRAW
Henny **COLENBRANDER**	COWAR
Kenneth A. **EDWARDS**	UN-DTCD
Rainer **ENDERLEIN**	UN-ECE
Cengiz **ERTUNA**	UN-ESCAP
Frank **HARTVELT**	UNDP
Richard **HELMER**	UNCED
Jorge **ILLUECA**	UNEP
Terence R. **LEE**	UN-ECLAC
Guy **LE MOIGNE**	World Bank
Peter N. **MWANZA**	UN-ECA
Pierre **NAJLIS**	UN-DIESA
Bede N. **OKIGBO**	UNU
Ahmad **RADJAI**	UN-ESCWA
Carel de **ROOY**	UNICEF
Alexander **ROTIVAL**	Water Supply and Sanitation Collaborative Council (until 1 October 1991)
Swayne **SCOTT**	FAO
Gehan **SINNATAMBY**	UNCHS-HABITAT
Andras **SZOLLOSI-NAGY**	UNESCO
Dennis **WARNER**	WHO
Ranjith **WIRASINHA**	Water Supply and Sanitation Collaborative Council (from 1 October 1991)
Yuecel **YURTSEVER**	IAEA

Annex 6

Acronym List

Institutions

UN Bodies and Agencies

DIESA	Department of International and Economic Affairs (UN)
DTCD	Department of Technical Co-operation for Development (UN)
ECA	Economic Commission for Africa (UN)
ECE	Economic Commission for Europe (UN)
ECLAC	Economic Commission for Latin America and Caribbean (UN)
ECOSOC	Economic and Social Council (UN)
ESCAP	Economic and Social Commission for Asia and Pacific (UN)
ESCWA	Economic and Social Commission for Western Asia (UN)
FAO	Food and Agriculture Organization (UN)
IAEA	International Atomic Energy Agency (UN)
IBRD	International Bank for Reconstruction and Development
ILO	International Labor Organization
PROWWESS	Promotion of the Role of Women in Water and Environmental Sanitation Services
RAPA	Regional Office for Asia and the Pacific (FAO)
ROSTA	Regional Office for Science and Technology – Africa (UNESCO)
ROSTAS	Regional Office for Science and Technology – Arab States (UNESCO)
ROSTCA	Regional Office for Science and Technology – South and Central Asia (UNESCO)
ROSTEA	Regional Office for Science and Technology – South and East Asia (UNESCO)
ROSTLAC	Regional Office for Science and Technology – Latin America and the Caribbean (UNESCO)
UN	United Nations
UNCED	United Nations Conference on Environment and Development (UN)
UNDP	United Nations Development Programme
UNDTCD	United Nations Department of Technical Cooperation for Development
UNEP	United Nations Environment Programme
UNESCO	United Nations Education, Scientific and Cultural Organization
UNFPA	United Nations Fund for Population Activities
WB	World Bank
WFP	World Food Programme

WHO	World Health Organization
WMO	World Meteorological Organization

UN Programmes and Commissions

ACC-ISGWR	Administrative Committee on Co-ordination – Intersecretariat Group on Water Resources (UN)
CDP	Committee on Development Planning (UN)
CEHA	Centre for Environmental Health Activities (WHO)
CHY	Commission for Hydrology (WMO)
CNR	Committee on Natural Resources (UN)
DGIP	Division for Global and Interregional Programmes
FREND	Flow Regimes from Experimental and Network Data (UNESCO)
GEMS/WATER	Global Freshwater Quality Monitoring Programme (UNEP/WHO/UNESCO/WMO)
GEMS	Global Environmental Monitoring System (UNEP)
HNRC	HOMS National Reference Centre (WMO)
HOMS	Hydrological Operational Multipurpose System (WMO)
HWRP	Hydrology and Water Resources Programme (WMO)
IAP-WASAD	International Action Programme on Water and Sustainable Agricultural Development
ICWE	International Conference for Water and the Environment
IDNDR	International Decade for Natural Disaster Reduction (UN)
IDWSSD	International Drinking Water Supply and Sanitation Decade (WHO)
IFAD	International Fund for Agricultural Development
IGADD	Intergovernmental Authority on Drought and Development
IHP/NC	International Hydrological Programme – National Committees (UNESCO)
IHP	International Hydrological Programme (UNESCO)
INFOHYDRO	Hydrological Information Referral Service (WMO)
ISGWR	Intersecretariat Group for Water Resources
MPAP	Mar del Plata Action Plan (UN)
OHP	Operational Hydrology Programme (WMO)
PEEM	Panel of Experts for Environmental Management for Vector Control (WHO/FAO/UNEP)
TCP	Tropical Cyclone Programme (WMO)
VCP	Voluntary Co-operation Programme (WMO)
WCP	World Climate Programme (WMO)
WWW	World Weather Watch Programme (WMO)

Intergovernmental Organizations

ABD	Asian Development Bank
ASEAN	Association of South East Asian Nations
GRDC	Global Run-off Data Centre (WMO/Federal Republic of Germany)
IDB	Inter-American Development Bank
OECD	Organisation for Economic Co-operation and Development
SADCC	Southern African Development Co-ordination Conference

Non-governmental Organizations

IAH	International Association of Hydrogeologists
IAHS	International Association of Hydrological Sciences
ICID	International Commission on Irrigation and Drainage
ICSU	International Council of Scientific Unions
IDA	International Development Association
IGBP	International Geosphere Biosphere Programme (ICSU)
IIMI	International Irrigation Management Institute
IRRI	International Rice Research Institute
ITN	International Training Network for Water and Waste Management
IWRA	International Water Resources Association

Nationally based international organizations

CEFIGRE	Centre de Formation Internationale pour la Gestion des Resources en Eau
DGIS	Directorate General for International Cooperation, The Netherlands
EDI	Economic Development Institute
EIER	Ecol Inter-Etats des Ingénieurs de l'Equipment Rural
IHE	Infrastructure Hydraulics Environmental
IRC	International Resources Center, The Netherlands
ORSTOM	Institut Francais de Recherche Scientifique pour le Developement en Cooperation
WASH	Water and Sanitation for Health
WEDC	Water Engineering and Development Centre
WSTC	Water Supply Training Center

Technical terms

ARIS	Annual Review of Implementation and Supervision
CB	Capacity Building
ENSO	El Niño Southern Oscillation
ESA	External Support Agency
GIS	Geographic Information System
HRD	Human Resources Development
ID	Institutional Development
R & D	Research and Development
UFW	Unaccounted For Water
WRA	Water Resources Assessment
WSS	Water Supply and Sanitation

References

Ambio: A Journal of the Human Environment, 1992. *Special Issue on Population, Natural Resources and Development.* Royal Swedish Academy of Sciences. Vol. XXI, No. 1, Feb., 1992.

Ayibotele, N.B., 1992. *The World's Water.* Keynote paper, International Conference on Water and the Environment, Dublin, Ireland, 26–31 Jan. 1992. 26 pp.

Ayibotele, N.B. and Falkenmark, M., 1992. Freshwater Resources. Chapter 10 (pp.187–203) in *An Agenda for Science for Environment and Development into the 21st Century.* Cambridge University Press.

Bamberger, M. and Cheema, S., 1990. *Case Studies of Project Sustainability.* Economic Development Institute of the World Bank.

Bernthal, F., (ed.), 1990. *Climate Change: The IPCC Response Strategies.* WMO-UNEP, Geneva.

Bhatia, R. and Falkenmark, M., 1992. *Water Resource Policies and the Urban Poor. Innovative Approaches and Policy Imperatives.* Background paper, International Conference on Water and the Environment, Dublin, Ireland, 26–31 Jan. 1992. World Bank.

Borde, J-P. and Pearce, D.W., 1991. *Valuing the Environment. Six Case Studies.* Earthscan Publications, London.

ECE, UN-ECOSOC, 1991. *Seminar on Ecosystems Approach to Water Management.* Report of the Seminar. Oslo, May, 1991.

ESCAP, 1991. *Report of the Workshop on Sustainable and Environmentally Sound Development of Water Resources.* Bangkok, 28 Oct.–1 Nov. 1991. Strategy document, International Conference on Water and the Environment, Dublin, Ireland, 26–31 Jan. 1992. 17 pp.

Falkenmark, M. and Lundqvist, J., 1992. *Coping with Multi-cause Environmental Challenges – a Water Perspective on Development.* Keynote paper, International Conference on Water and the Environment, Dublin, Ireland, 26–31 Jan. 1992. 22 pp.

Falkenmark, M, Lundqvist, J. and Widstrand, C., 1990. *Water Scarcity – an Ultimate Constraint in Third World Development. A Reader on a Forgotten Dimension in Dry Climate Tropics and Subtropics.* Dept. of Water and Environmental Studies, University of Linkoping, Sweden.

FAO, 1990. *An International Action Programme on Water and Sustainable Agricultural Development.* FAO, Rome.

FAO, 1991(a). *Water and Sustainable Agricultural Development.* Strategy document, International Conference on Water and the Environment, Dublin, Ireland, 26–31 Jan. 1992. 42 pp.

FAO, 1991(b). *Agricultural Water Use: Assessment of Progress in the Implementation of the Mar del Plata Action Plan.* Strategy document, International Conference on Water and the Environment, Dublin, Ireland, 26–31 Jan. 1992. 74 pp.
FAO/WHO/UNDP/UNICEF, 1991. *Water for Sustainable Food Production and Drinking Water Supply and Sanitation in the Rural Context.* Background paper, International Conference on Water and the Environment, Dublin, Ireland, 26–31 Jan. 1992. 14 pp. + worksheets.
Gupta, D.B., 1992. *The Importance of Water Resources for Urban Socio-economic Development.* Keynote paper, International Conference on Water and the Environment, Dublin, Ireland, 26–31 Jan. 1992. 19 pp.
Horchani, A, 1992. *Environmental Health Issues: Impacts of Water and Waste Management.* Keynote paper, International Conference on Water and the Environment, Dublin, Ireland, 26–31 Jan. 1992. 9 pp.
Houghton, J.T., Jenkins, G.J. and Ephraums, J.J. (eds.), 1990. *Climate Change: the IPCC Scientific Assessment.* Cambridge University Press, 364 pp.
ICWE, 1992. *Report of the Conference.* International Conference on Water and the Environment, Dublin 1992.
IHE-Delft/UNDP, 1991. *A Strategy for Water Sector Capacity Building.* Strategy document, International Conference on Water and the Environment, Dublin, Ireland, 26–31 Jan. 1992. 191 pp.
IUCN/UNEP/WWF, 1991. *Caring for the Earth: A Strategy for Sustainable Living.* 228p.
Jäger, J. and Ferguson, H.L., 1991. *Climate Change: Science, Impacts and Policy.* Proceedings of the Second World Climate Conference. Cambridge University Press, 578 pp.
Kalinen, G.P. and Bykov, V.D., 1969. The World's water resources, present and future. From *Impact of Science on Society*, Vol. XIX, No. 2, 1969, UNESCO. In *The Ecology of Man: An Ecosystem Approach*, ed. R.L. Smith, pp. 335–346. Harper and Row, New York.
Kindler, J., 1992. *Water – the Environmental and Developmental Dimensions -Striking a Balance.* Keynote paper, International Conference on Water and the Environment, Dublin, Ireand, 26–31 Jan. 1992. 19 pp.
Korzun, V.I. (ed.), 1974. *World Water Balance and Water Resources of the Earth.* (Translated from Russian by UNESCO, 1978, Paris, France, 664 pp.)
Koudstaal, R., Rijsberman, F.R. and Savenije, H., 1992. *Water and Sustainable Development.* Keynote paper, International Conference on Water and the Environment, Dublin, Ireland, 26–31 Jan. 1992. 23 pp.
L'vovich, M.I., 1979. *World Water Resources and their Future.* (In Russian; English edition edited by R. L. Nace, 1979, *American Geophysical Union*, Washington, DC. 415 pp).
Meybeck, M., Chapman, D. and Helmer, R., (eds.) 1989. *Global Freshwater Quality: a First Assessment.* WHO/UNEP. Blackwell, Oxford.
Nace, R.L., 1969. World water inventory and control. In *Water, Earth and Man*, ed. R.J. Chorley, Methuen, London. 588 pp.
National Research Council, 1991. *Opportunities in the Hydrological Sciences.* National Academy Press, Washington.
OAS, 1984. *Integrated regional development planning guidelines and case studies from OAS experience.* Department of Regional Development, Organisation for American States. January, 1984.
Pearce, D. and Markandya, A., 1989. *Environmental Policy Benefits: Monetary Valuation.* OECD, Paris, 1989.
Peters, A., 1990. *Peters' Atlas of the World.* Longman.

Plate, E.J., 1992. *Scientific and Technological Challenges*. Keynote paper, International Conference on Water and the Environment, Dublin, Ireland, 26–31 Jan. 1992. 21 pp.

Rodda, A., 1991. *Women and the Environment.* United Nations Non-governmental Liaison Service. Zed Books Ltd., 180 pp.

Rogers, P, 1992. *Integrated Urban Water Resources Management*. Keynote paper, International Conference on Water and the Environment, Dublin, Ireland, 26–31 Jan. 1992. 39 pp.

Rydzewski, J.R. and Abdullah, S. bin, 1992. *Water for Sustainable Food and Agricultural production.* Keynote paper, International Conference on Water and the Environment, Dublin, Ireland, 26–31 Jan. 1992. 25 pp.

Sayre, I.A., 1988. *International Standards for Drinking Water*. Journal of the AWWA, January 1988, 53–60.

Shiklomanov, I.A., 1991. The world's water resources. In *International Symposium to Commemorate the 25 years of IHD/IHP*. Unesco, Paris.

Snead, R. E., 1972. *Atlas of World Physical Features.* Wiley.

Strahler, A. N. and Strahler, A. H., 1987. *Modern Physical Geography*. 3rd edn. John Wiley & Sons, Toronto.

Tegart, W.J. McG., Sheldon, G.W. and Griffiths, D.C., 1990. *Climate Change: the IPCC Impact Assessment*. Australian Government Publishing Service, Canberra.

Traoré, A., 1992. *Water for the People – Community Water Supply and Sanitation.* Keynote paper, International Conference on Water and the Environment, Dublin, Ireland, 26–31 Jan. 1992. 22 pp.

UN, 1993. *Earth Summit Agenda 21, the UN Programme of Action from Rio*. Text of the United Nations Conference on Environment and Development (UNCED), 3–14 June 1992, Rio de Janeiro, Brazil, 294 pp.

UN-DIESA, 1991(a). *World Population Prospects* 1990. Population Studies No. 120, New York.

UN-DIESA, 1991(b). *Current Trends and Policies in the World Economy.* World Economic Survey 1991. New York.

UN-DIESA/UN-DTCD/UNDP/IBRD/UNCED, 1991. *Mechanisms for Implementation and Co-ordination at Global, National, Regional and Local Levels.* Background Paper International Conference on Water and the Environment, Dublin, Ireland, 26–31 Jan. 1992. 15 pp. + worksheets.

UN-DTCD, 1991(a). *Demand Management*. Strategy document, International Conference on Water and the Environment, Dublin, Ireland, 26–31 Jan. 1992. 65 pp.

UN-DTCD, 1991(b). *Integrated Water Resources Planning.* Strategy document, International Conference on Water and the Environment, Dublin, Ireland, 26–31 Jan. 1992. 128 pp.

UN-DTCD, 1991(c). *Water Management.* Strategy document, International Conference on Water and the Environment, Dublin, Ireland, 26–31 Jan. 1992. 33 pp.

UN-DTCD, 1991(d).Working paper on implementation mechanisms for integrated water resources development and management. Copenhagen Statement on Integrated Water Resources Management. *Copenhagen Informal Consultation, Nov. 1991.* Strategy document, International Conference on Water and the Environment, Dublin, Ireland, 26–31 Jan. 1992.

UN-DTCD/IBRD/UNDP, 1991. *Integrated water resources development and management.* Background Paper, International Conference on Water and the Environment, Dublin, Ireland, 26–31 Jan. 1992. 12 pp. + worksheets.

UNCHS-HABITAT/WHO/IBRD/UNDP/UNICEF, 1991. *Water for Sustainable Urban Development and Drinking Water Supply and Sanitation in the Urban Context.*

Background paper, International Conference on Water and the Environment, Dublin, Ireland, 26–31 Jan. 1992. 17 pp. + worksheets.

UNDP, 1991. *Safe Water 2000: Report of the Global Consultation on Safe Water and Sanitation for the 1990s.* Strategy document, International Conference on Water and the Environment, Dublin, Ireland, 26–31 Jan. 1992. 76 pp.

UNEP, 1991. *Freshwater Pollution.* Strategy document, International Conference on Water and the Environment, Dublin, Ireland, 26–31 Jan. 1992. 36 pp.

UNEP/WHO, 1991. *Protection of water resources, water quality and aquatic ecosystems.* Background paper, International Conference on Water and the Environment, Dublin, Ireland, 26–31 Jan. 1992. 10 pp. + worksheets.

United Nations, 1977. *Report of the United Nations Water Conference, Mar del Plata, 14 --25 March 1977.* E/CONF.70/29. 138 pp. and 2 annexes.

United Nations, 1980. *Efficiency and Distributional Equity in the Use and Treatment of Water: Guidelines for Pricing and Regulation.* DIESA/DTCD, Natural Resources/Water Series No. 8, New York.

Water International, 1991(a). *Special Issue on International Drinking Water Supply and Sanitation Decade.* Vol. 16, No. 3, Sept. 1991.

Water International, 1991(b). IWRA Statement on Water, Environment and Development. Vol. 16, No. 4, 1991, 243–246.

WHO, 1991. *The International Drinking Water Supply and Sanitation Decade: Review of Decade Progress (as at December 1990).* Division of environmental Health, WHO, Geneva.

WHO/UNEP, 1991. *Water Quality.* Strategy document, International Conference on Water and the Environment, Dublin, Ireland, 26–31 Jan. 1992. 79 pp.

Williams, W. D., 1987. *Ambio*, **16**, 180–185.

WMO, 1992. *Statement and Report, Dublin International Conference on Water and the Environment – Development Issues for the 21st Century*, 55 pp.

WMO/UNEP, 1990(a). *Climate Change. The IPCC Scientific Assessment.* Intergovernmental Panel on Climate Change. Cambridge University Press, 365 pp.

WMO/UNEP, 1990(b). *Climate Change. The IPCC Impacts Assessment.* Intergovernmental Panel on Climate Change.

WMO/UNEP, 1990(c). *Climate Change. The IPCC Response Strategies.* Intergovernmental Panel on Climate Change, 270 pp.

WMO/UNEP, 1992. *Climate Change 1992. The Supplementary Report to the IPCC Scientific Assessment.* Intergovernmental Panel on Climate Change. Cambridge University Press, 200 pp

WMO/UNESCO, 1991(a). *Water Resources Assessment.* Strategy document, International Conference on Water and the Environment, Dublin, Ireland, 26–31 Jan. 1992. 64 pp.

WMO/UNESCO, 1991(b). *Water resources assessment and impacts of climate change on water resources.* Background Paper, International Conference on Water and the Environment, Dublin, Ireland, 26–31 Jan. 1992. 13 pp. + worksheets.

WMO/WHO/UNEP, 1991. *Information Needs for Water Quality Assessment and Management. Report of an expert consultation (Bratislava, 26–30 Aug. 1991).* Strategy document, International Conference on Water and the Environment, Dublin, Ireland, 26–31 Jan. 1992. 18 pp. + annexes.

World Commission on Environment and Development, 1987. *Our Common Future.* Oxford University Press.

World Resources Institute, 1990. *World Resources 1990–91: A Guide to the Global Environment.* Oxford University Press.

Index

Printed in the United States
By Bookmasters